BY SUSAN GREENFIELD

MIND CHANGE

MIND
CHANGE

HOW DIGITAL
TECHNOLOGIES
ARE LEAVING
THEIR MARK ON
OUR BRAINS

SUSAN GREENFIELD

RANDOM HOUSE NEW YORK

Published in the United States by Random House, an imprint and division of Random House LLC, a Penguin Random House Company, New York.

RANDOM HOUSE and the HOUSE colophon are registered trademarks of Random House LLC.

Extract taken from "The Horses" taken from *Collected Poems* © Estate of Edwin Muir and reprinted by permission of Faber and Faber Ltd.

Library of Congress Cataloging-in-Publication Data
Greenfield, Susan.
Mind change : how digital technologies are leaving their mark on our brains / Susan Greenfield.—First edition.
pages cm
Includes index.
ISBN 978-0-8129-9382-0 (hardcover)—ISBN 978-0-8129-9383-7 (eBook)
1. Cognition. 2. Information technology—Psychological aspects.
3. Information technology—Social aspects. I. Title.
BF311.G7135 2015
155.9—dc23
2014020059

Printed in the United States of America on acid-free paper

www.atrandom.com

2 4 6 8 9 7 5 3 1

First Edition

Book design by Liz Cosgrove

For John—invaluable friend, mentor, and colleague

CONTENTS

PREFACE

The events leading up to the writing of *Mind Change* have been unfolding for the past five years, and arguably for much longer than that—perhaps unknowingly ever since I started neuroscience research and began to realize the power and vulnerability of the human brain. True, my main focus over several decades has been trying to uncover the basic neuronal mechanisms accountable for dementia, literally a loss of the mind. But even before I ever put on a white coat, it was the still broader and more general question of what might be the physical basis for the mind itself that held an utter fascination for me. Having made the rather unconventional journey to brain research from classics via philosophy, I was always interested in the big questions of whether we truly have free will, how the physical brain might generate the subjective experience of consciousness, and what makes every human being so unique.

Once I was in the lab, some aspects of these tantalizing issues could be translated into specific questions that might be tested experimentally. Accordingly, over the years, we've researched the impact of a stimulating, interactive "enriched" environment on brain processes, as well as the release and action of the versatile and hard-working chemical messenger dopamine, in turn linked to the subjective experiences of reward, pleasure, and addiction. At a more applied level, we've investigated how the drug Ritalin, used to treat attention deficit hyperactivity disorder (ADHD), might work, and how insights from neuroscience can contribute to improved performance in the classroom. Yet there has always been an underlying theme common to all these diverse areas of inquiry, including our research on neurodegenerative disorders: novel brain mechanisms, how they might be inappropriately activated in disease, and, more generally, how these as yet underappreciated neuronal processes enable each of us to adapt to our own individual environment—to become individuals.

This wonderful *plasticity* of the human brain served as a natural segue into thinking about the future and how upcoming generations might adapt to the very different, highly technological landscape of the decades to come. Accordingly, in 2003 I wrote *Tomorrow's People,* exploring the possible new kinds of environment and lifestyle that information technology, biotechnology, and nanotechnology in combination would deliver. In turn, this very different potential world led me to reflect more on the implications for identity. In 2007, these ideas were set out in *ID: The Quest for Meaning in the 21st Century,* which subsequently was to inspire a novel of a dystopian future (*2121*). In *ID,* I had suggested that three broad options had historically presented themselves for self-expression. The *somebody* scenario of defining oneself via consumerism offered an individual identity without true fulfillment, while the *anyone* alternative of a collective identity resulted in the opposite, fulfillment that was subsumed into a wider impersonal narrative. Finally, there was the *nobody* possibility traditionally achieved with wine, women, and song, where the sense of self is abnegated in favor of being a passive recipient of the incoming senses. When you have a "sensational" time, I argued, you were no longer "self"-conscious.

But might the supra-sensational digital technologies of the twenty-first century be shifting the balance away from an occasional and contrived situation (drinking, fast-paced sports, dancing) in favor of the "mind-blowing" scenario becoming more the default cognitive mode? These thoughts were in the background of my mind when, in February 2009, I had the chance to articulate them more clearly.

There was a debate in the United Kingdom's House of Lords on the regulation of websites, particularly with regard to children's well-being and safety. If you sign up to speak at such an event, the convention is that you present an argument based on your own specific area of expertise. Since I knew nothing whatsoever about legislation and regulatory practice, I decided to offer a perspective through the prism of neuroscience. The syllogism I used was quite straightforward and not particularly original. Any neuroscientist might well have said the same thing: the human brain adapts to the environment and the environment is changing in an unprecedented way, so the brain may also be changing in an unprecedented way.

The reaction by the international print and broadcast media to this seemingly bland and logical argument was out of all proportion to its content. Needless to say, I had to endure the inevitable press misrepresentation resulting from a priority of selling copy over the actual truth: "Baroness Says Computers Rot the Brain" was just one of the more lurid headlines. Meanwhile, I was also told by journalists interviewing me, with the glee that people have for imparting bad news, how reviled I was in some quarters of the blogosphere, and then asked how I felt about it.

My reaction was, and has been, that I'm happy to discuss the science prompting my ideas and that I will wave the white flag if trumped by hard facts. That is what scientists do: it's how we publish our peer-reviewed papers and it's how we develop theories. Most of us take professional criticism as the warp and weft of the research process. However, what was really interesting here was the apparent ferocity of the personalized animosity in some cases. Had I said the earth was flat, I doubt if anyone would have cared. Clearly I was touching a very raw nerve that made some people feel threatened or in some way undermined. Until then, I hadn't realized just how important an issue this was for our society. I therefore continued to read more, to think, and to speak in a wide range of forums about the brain of the future—indeed, the future of the brain.

Then, on December 5, 2011, the House of Lords presented a further opportunity for more formal open discussion. I had the chance to introduce a debate to "ask Her Majesty's Government what assessment they have made of the impact of digital technologies on the mind." As you can imagine, securing parliamentary time in the historic red and gold chamber is not easy, and I felt very fortunate even to have been given the brief slot that is known as a "Question for Short Debate." Present at this debate were a range of representatives from diverse sectors, ranging from business to education to medicine.

Interestingly enough, most of the noble lords seemed keen to emphasize the benefits of technology, and the general tone from the majority of those speaking gave the impression that there was no need for immediate concern. In his summing up, the then British parliamentary undersecretary of state for schools, Lord Hill of Oareford, concluded that he was "not aware of an extensive evidence base on negative impact from the sensible and proportionate use of technology," although, "just as any technological revolution can lead to great progress, so it always also leads to unexpected problems, to which we must indeed always be alert."

One of the drawbacks of the format of Questions for Short Debate is that, as the name suggests, time is short and, unlike with lengthier slots of different types, the peer who has instigated the particular topic, as I had done on this occasion, is unable to reply to the ideas that have been aired. Needless to say, if I had had the chance, I would have questioned the minister on four basic points.

First, very little is currently being done by the U.K. government to promote research into the effects of screen culture on the young mind, or indeed the mind at any age. If such an initiative was undertaken, it would be vital to know what kind of research was being done, in what kind of areas, how much funding was being provided, and over what period of time they were anticipating conducting these studies.

My second point would have been that if technology is indeed being used "sensibly" (in itself a subjective judgment), then by definition such "sensible" practices could not have a significantly negative impact. The whole point I had been trying to make was that technology is not necessarily being used in moderation; some surveys have suggested that it is used up to eleven hours a day. Would this really qualify as "proportionate"?

My third point would have been that when we look at the various aspects of cyberculture, there is indeed reason for concern. Yet this ministerial speech serves as a good example of a strategy popular not just with politicians and civil servants but also with anyone wanting a quiet life: prevarication until more evidence comes in, without any indication of just how much and what type of evidence would be convincing enough to launch a wide discussion involving policy makers, parents, teachers, and taxpayers more generally. So my fourth and final point would have been that the unspecified "problems" mentioned by the minister will be "unexpected" only if we neither anticipate nor discuss them.

And just at that very moment, with the uncanny coincidences that can sometimes occur in life, I was approached by Random House to write this book. *Mind Change* could therefore be viewed in one sense as an answer to the minister, but its main purpose is to meet the needs of a society that should be squaring up to make some decisions. In order to do so, we must have a balanced and comprehensive overview of the scientific research. While such an overview can never be exhaustive, it should cover the most significant findings—and that is what you will find in this book. It's worth noting, however, that one deliberate omission is the field of Internet pornography, where the controversy and debate are obviously not so much about whether it is "good" or "bad," or about how it impacts on types of thinking, but more about legislation and regulation, which are outside the scope of the journey here.

The main goal of *Mind Change* is to explore the different ways in which digital technologies could be affecting not just thinking patterns and other cognitive skills but also lifestyle, culture, and personal aspirations. Accordingly, in addition to coverage of the peer-reviewed scientific literature, you'll find discussion of the various goods and services that could be revealing a new type of mindset, as well as commentaries and reports in the popular press that act as a mirror to the society in which we live.

Discovering and collating such a wide range of diverse types of material would be—indeed, *is*—extremely daunting. However, once again fate took a hand, and at a beach party in Melbourne in December 2012, I was fortunate enough to meet Olivia Metcalf. Olivia had just finished a PhD at the Australian National University in Canberra, studying videogames, and was unsure of the path she wished to take. Amazingly, she was available and willing to help ensure that the manuscript, then in first draft, encompassed the wide range of research into digital technologies. Over the subsequent year, Olivia's contribution has been invaluable. Her scrutiny and critique of the work have truly raised the game of what *Mind Change* can offer: an in-depth perspective on a highly complex and fast-moving field.

Some thirty-five years ago, while I was working in Paris, a colleague showed me the front page of a newspaper featuring a heavily bearded man in a sweater of dubious taste. "He's from the green movement," he sneered, laughing at the individual as a weird eccentric. The idea of a "green" movement certainly seemed strange to me, as did the phrase "climate change." Now this concept significantly impinges on much of public policy and influences individual lifestyle. *Mind Change* is so called because I suggest that there are similar parallels with climate change, albeit lagging behind by several decades: both are global, controversial, unprecedented, and multifaceted. While the challenges of climate change require exercises in damage limitation, *Mind Change* could open up the most exciting possibilities for twenty-first-century society to realize the full potential of each human mind as never before, if only we can discuss and plan what kind of world we want to live in or, more specifically, what kind of people we actually want to be.

MIND CHANGE

1

MIND CHANGE

A GLOBAL PHENOMENON

Let's enter a world unimaginable even a few decades ago, one like no other in human history. It's a two-dimensional world of only sight and sound, offering instant information, connected identity, and the opportunity for here-and-now experiences so vivid and mesmerizing that they can outcompete the dreary reality around us. It's a world teeming with so many facts and opinions that there will never be enough time to evaluate and understand even the smallest fraction of them. For an increasing number of its inhabitants, this virtual world can seem more immediate and significant than the smelly, tasty, touchy 3-D counterpart: it's a place of nagging anxiety or triumphant exhilaration as you are swept along in a social networking swirl of collective consciousness. It's a parallel world where you can be on the move in the real world, yet always hooked into an alternative time and place. The subsequent transformation of how we might all be living very soon is a vitally important issue, perhaps even *the* most important issue of our time.[1] Why? Because it may be that a daily existence revolving around smartphone, iPad, laptop, and Xbox is radically changing not just our everyday lifestyles but also our identities and even our inner thoughts in unprecedented ways.[2] As a neuroscientist, I'm fascinated by the potential effects of a screen-oriented daily existence on how we think and what we feel, and I want to explore how that exquisitely adaptable organ, the brain, may now be reacting to this novel environment, recently dubbed the "digital wildfire."[3]

In the developed world, there is now a one in three chance that children will live to 100 years of age.[4] Thanks to the advances of biomedicine, we can anticipate longer and healthier lives; thanks to technology, we can foresee an existence increasingly freed from the daily domestic grind that characterized the lives of previous generations. Unlike so much of humanity in the past and still in many nightmare scenarios around the world, we take it as the norm and as our entitlement not to be hungry, cold, in pain, or in constant fear for our lives. Unsurprisingly, therefore, there are many in our society who are convinced that we're doing just fine, that these digital technologies are not so much a raging wildfire but more of a welcoming hearth at the heart of our current lifestyles. Accordingly, various reassuring arguments are ready at hand to counter reservations and concerns that might otherwise be viewed as exaggerated, even hysterical.

One starting premise is that surely everyone has enough common sense to ensure that we don't let the new cyberculture hijack daily life wholesale. Surely we are sensible and responsible enough to self-regulate how much time we spend online and to ensure that our children don't become completely obsessed by the screen. But the argument that we are automatically rational beings does not stand the test of history: when has common sense ever automatically prevailed over easy, profitable, or enjoyable possibilities? Just look at the persistence of hundreds of millions worldwide who still spend money on a habit that caused a hundred million fatalities in the twentieth century and which, if present trends continue, promises up to one billion deaths in this century: smoking.[5] Not much common sense at work there.

Then again, the reliability of human nature might work in our favor if only we could assume that our innate genetic makeup leads most of us to do the right thing, regardless of any corrupting external influences. Yet in itself, this idea immediately runs counter to the superlative adaptability of the human brain, which allows us to occupy more ecological niches than any other species on the planet. The Internet was initially created as a way for scientists to contact each other, and this invention spawned phenomena such as 4chan, a collection of message boards where people post images and short text comments, mostly anonymously and with no holds barred.[6] This form of self-expression is a new niche to which we may adapt, with consequences as extreme as the medium itself. If it is the hallmark of our species to thrive wherever we find ourselves, then the digital technologies could bring out the worst in human nature rather than being rendered harmless by it.

Another way of dismissing out of hand the concerns that the effects of digital technology may bring is a kind of solipsistic stance in which the screen enthusiast proudly points to his or her own perfectly balanced existence, which combines the pleasures and advantages of cyberculture with life in three dimensions. Yet psychologists have been telling us for many years that such subjective introspection is an unreliable barometer of mental state.[7] In any case, it should be obvious enough that just because a single individual may be able to achieve an ideal mix between the virtual and the real, it does not automatically mean that others are capable of exercising similar restraint and sound judgment. And even those individuals who think they've got everything just right will often admit in an unguarded moment that "It's easy to waste a lot of time on Facebook," that they are "addicted" to Twitter, or that, yes, they do find it hard to concentrate long enough to read a whole newspaper article. In the United Kingdom, the advent of *I,* an abbreviated version of the national quality paper *The Independent,* and the introduction on the BBC of the *90 Second News Update* stand as testimony to the demands of an ever larger constituency of readers and viewers—not just the younger generation—who have a reduced attention span and are demanding print and broadcast media to match.

Another consolation is the conviction that the next generation will work out just fine, thanks to parents who take control and intervene where necessary. Sadly, this idea has already proved to be a nonstarter. For reasons we shall explore shortly, parents often complain that they cannot control what their offspring do online, and many already despair at their inability to prize their children away from the screen and back into a world of three dimensions.

Marc Prensky, an American technologist, coined the term "Digital Native" for someone defined by his or her perceived outlook and abilities, based on an automatic facility and familiarity with digital technologies.[8] By contrast, "Digital Immigrants" are those of us who, according to Prensky, "have adopted many aspects of the technology, but just like those who learn another language later in life, retain an 'accent' because we still have one foot in the past." It is unlikely that anyone reading these words will not have strong views as to which side of the divide he or she belongs and whether the distinction is cause for unalloyed celebration or deep anxiety. Generally speaking, it corresponds to age, although Prensky himself did not pinpoint a specific line of demarcation. The date of birth of the Digital Native seems therefore to be uncertain: we could start as far back as the 1960s, when the term "computer" entered into common parlance, or as late as 1990, for by the time a young Digital Native born then could read and write, email (which started around 1993) would have become an inescapable part of life.

The important distinction is that Digital Natives know no other way of life other than the culture of Internet, laptop, and mobile. They can be freed from the constraints of local mores and hierarchical authority and, as autonomous citizens of the world, will personalize screen-based activities and services while collaborating with, and contributing to, global social networks and information sources.

But a much gloomier portrait of the Digital Native is being painted by pundits such as the British American author Andrew Keen:

> MySpace and Facebook are creating a youth culture of digital narcissism; open-source knowledge sharing sites like Wikipedia are undermining the authority of teachers in the classroom; the YouTube generation are more interested in self-expression than in learning about the world; the cacophony of anonymous blogs and user-generated content is deafening today's youth to the voices of informed experts.[9]

Then again, perhaps the Digital Native doesn't actually exist after all. Neil Selwyn, of the Institute of Education in London, argues that the current generation is actually no different from preceding ones: young people are not hardwired to have unprecedented brains.[10] Rather, many young people are using technology in a far more sporadic, passive, solitary, and, above all, unspectacular way than the hype of the blogosphere and zealous proponents of cyberculture might have us believe.

Irrespective of whether the digital age has spawned a new type of superbeing or just ordinary humans better adapted to screen life, suffice it to say that, for the moment, parents are most likely to be Digital Immigrants and their children Digital Natives. The former are still learning the enormous potential of these technologies in adulthood, while the latter have known nothing else. This cultural divide often makes it hard for parents to know how best to approach situations that they intuitively perceive to be a problem, such as seemingly excessive time spent on computer-based activities; meanwhile, children may feel misunderstood and impatient with views they regard as inappropriate and outdated for present-day life.

Although reports and surveys have focused largely on the next generation, the concerns I want to flag are not limited to the Digital Native alone. Far from it. But a generational divide has undoubtedly arisen from the vertiginous increase in the pace of ever smarter digital devices and applications. What will be the effects on each generation, and on the relationship between them?

In a 2011 report, *Virtual Lives,* researchers for the U.K. children's charity Kidscape assessed the online activities of more than two thousand children between the ages of eleven and eighteen. Just under half of the children questioned said they behaved differently online compared to their normal lives, with many claiming it made them feel more powerful and confident. One explained: "It's easier to be who you want to be, because nobody knows you and if you don't like the situation you can just exit and it is over." Another echoed this sentiment, noting: "You can say anything online. You can talk to people that you don't normally speak to and you can edit your pictures so you look better. It is as if you are a completely different person." These findings, the report argues, "suggest that children see cyberspace as detachable from the real world and as a place where they can explore parts of their behavior and personality that they possibly would not show in real life. They seem unable to understand that actions online can have repercussions in the real world."[11] The easy opportunity of alternative identity and the notion that actions don't have consequences have never previously featured in a child's development, and they are posing unprecedented questions as to what might be for the best. While the brain is indeed not hardwired to interface effectively with screen technologies, it has evolved to respond with exquisite sensitivity to external influences—to the environment it inhabits. And the digital environment is getting ever more pervasive at an ever younger age. Recently Fischer-Price introduced a potty-training seat complete with an iPad holder,[12] presumably to complement an infant lifestyle where the recliner in which the baby may spend many hours is also dominated by a screen.[13]

This is why the question of the impact of digital technologies is so very important. Hardened captains of industry or slick entrepreneurs will often sidle up to me during the coffee break at corporate events and let their professional mask slip as they recount in despair the obsessional fixation of their teenage son or daughter with the computer. But these anxieties remain unchanneled and unfocused. Where can these troubled parents share their experiences with others on a wider platform and articulate them in a formal and cogent way? At the moment, nowhere. In the following pages, we'll be looking at many studies on preteens as well as teenagers; unfortunately, there are far fewer studies on adults, perhaps because they are less cohesive and identifiable as a group than a volunteer student body or a captive classroom. But, in any event, it's important to view the data *not* as a self-help guide for bringing up kids but rather as a pivotal factor in the bigger picture of society as a whole.

Another argument sometimes used to dismiss any concerns about digital culture is the idea that we'll muddle through as long as appropriate regulation is in place. All too often we hear something like this from professional policy makers and government officials: *There is no conclusive evidence for concern as yet. If and when there is, all the appropriate checks and balances will of course be duly put in place. In the meantime, as long as we are sensible and proportionate, we can enjoy and benefit from all the advantages of the cyberlife. Technology clearly brings us previously unimagined opportunities, and such advances will of course be balanced out by always being alert to potential negative impacts.*[14] Yet while moderation may well be the key, technology is *not* necessarily being used in moderation. Young people in the United States, on average, use entertainment media more than fifty-three hours per week.[15] When media multitasking, or using more than one medium at once, is taken into account, young people average nearly eleven hours' worth of entertainment media use per day—hardly moderate.

The deeper problem with seeing regulation as the "solution" is that it is always reactive. Regulatory procedures can only respond to, and then sweep up behind, some new event, discovery, or phenomenon in order to eliminate clear harm, as with junk food, air pollution, or, to use an Internet example, the sexual grooming of children or their access to extreme violence. But regulation always has to play catch-up: politicians and civil servants will always be leery about predictions because they are rightly aware they are spending taxpayers' or donors' money on what could be regarded as speculation. However much guidelines and laws may be needed for the obvious and immediate dangers of the cyberworld, they are inadequate to the task of looking forward, of imagining the best uses to which new technologies can be put. For that we need long-term imagination and bold thinking, qualities not necessarily associated nowadays with cash-strapped civil servants or politicians with an eye to imminent reelection and easy wins in the short term. And so it is up to the rest of us. Technology can be empowering and can help us shape more fulfilling lives, but only if we ourselves step up to the plate and help take on the task.

Digital technologies are eroding the age-old constraints of space and time. I'll always remember a speech by former U.S. president Bill Clinton that I attended in Aspen, Colorado, back in 2004, where he described how the history of civilization could be marked by three stages: isolation, interaction, and integration. *Isolation* characterized the segregation of the remote empires of the past, access to which even until the last century was intermittent, time-consuming, and hazardous. *Interaction,* as Clinton pointed out, subsequently proved to be both positive, in the form of trade, exchanges of ideas, and so on, and negative, with the increased facility and scale of warfare. But this century is perhaps exemplifying for the first time the realization of a massive *integration.*

And yet this idea, at least as a hypothetical scenario, is not that revolutionary. As long ago as 1950, the French philosopher and Jesuit priest Pierre Teilhard de Chardin developed the idea of globalized thought, an eventual scenario he dubbed the "noosphere."[16]

According to Teilhard de Chardin, the noosphere would emerge through, and be composed of, the interaction of human minds. As humanity progressed into more complex social networks, the noosphere would be elevated in awareness. Teilhard de Chardin saw the ultimate apotheosis of the noosphere as the Omega Point, the greatest degree of collective consciousness to which the universe would evolve, with individuals still as distinct entities. Tempting as it is to believe that the digitally induced globalization in instant thought sharing and worldwide communication is realizing his vision, we cannot assume that this erstwhile hypothetical idea is now becoming our reality. What if one immediate outcome of global outreach and a correspondingly homogenized culture might be that we all start to react and behave in a more homogeneous style, one that eventually blurs cultural diversity and identity? Obviously, while there are huge advantages to understanding previously alien lifestyles and agendas, there is a big difference between a world enriched by other, contrasting ways of living and one that shares a single standardized, cookie-cutter existence. While diversity in societies brings great insights into the human condition, surely such comparisons can only be based on a clear and confident identity and lifestyle. A mere global homogenization of mindset might in the long run have serious consequences for how we see ourselves and the societies in which we live.

While speed, efficiency, and ubiquity must surely be good things, this new life of integration may have other, less beneficial effects that we need to think about. In days gone by we waited for the delivery of the mail at fixed times daily. An international phone call was, for everyone other than the very rich, generally an option only for special or emergency circumstances. But we now take for granted the constant availability of international communication. We tend to expect instant responses, and in turn assume we ourselves will reply immediately, oscillating incessantly between transmit and receive modes.

At a formal breakfast I attended recently where the main speaker was the British deputy prime minister, Nick Clegg, the woman sitting next to me was so busy tweeting that she was at a breakfast with Clegg that she wasn't actually listening to what he was saying. Twenty-four percent of users of U.S. adult social networking sites reported a curious phenomenon in 2012—that they missed out on a key event or moment in their lives because they were so absorbed in updating their social networking site about that event or moment.[17] Alternatively, you can monitor the flood of consciousness of others, almost as a way of life. When I asked a colleague how often she used Twitter, she showed me an email from a friend that is not uncommon in what it describes: "I have Twitter open on my PC all day so I look at it between calls, when on hold on the phone etc. I'd say pretty much our whole office does."

We no longer need to wait, to acknowledge the passing of time between cause and effect or between action and reaction. For most people who a few decades ago would never have contemplated foreign travel or having a network of friends beyond the local community into which they were born, there are now nonstop thrilling opportunities for encompassing the entire planet. The advantages of this effortless communication are many. No one could make a convincing case for turning back the clock to when postal deliveries took days. But perhaps there is some merit in having time to reflect before responding to views or information. Perhaps there are benefits to pacing your day according to your own choice, at your own speed.

The crucial issue here is how we digest internally what is happening around us as we travel through each day. The Austrian physician who developed the current treatment for Parkinson's disease back in the 1960s, Oleh Hornykiewicz, once offered this insight: "Thinking is movement confined to the brain." A movement is characterized by a chain of linked actions that take place in a particular order. The simplest example, walking, is a series of steps in which placing one foot forward leads to the other foot overtaking it; one step thus leads to the next in a cause-and-effect chain that is not random but a fixed linear sequence. So it is with thought. All thought, be it fantasy, memory, logical argument, business plan, hope, or grievance, shares this basic common characteristic of a fixed sequence. And since there is clearly a defined beginning, middle, and end to the sequence, there has to be a time frame. As I see it, this idea of sequence is the very quintessence of a thought, and it is the mental step needed that will distinguish a line or train of thought from a one-off instantaneous emotion captured in a shriek of laughter or a scream. Unlike a raw feeling that occurs as a momentary reaction, the thought process transcends the here and now and links past with future.

Human beings are not alone in possessing sufficient memory to link a previous event, a cause, with a subsequent one, an effect, and even to see a likely result in the future. A rat that receives a food pellet for pressing a bar can soon "think" about its next best move and learn to press the bar again. The link between stimulus and response has been forged. But we humans are unique in being able to link events, people, and objects that are *not* physically present in front of us into a stream of thought. We have the ability to see one thing, including an abstract word, in terms of another. Unlike all other animals, and even human infants, we have spoken and written language. We are liberated from the press of the moment around us because we can turn toward the past and then to the future by using symbols, words, to stand for things that are not physically present: we can remember and plan and imagine. But it takes time to do so, and the more complex the thought, the more time we need to take the necessary mental steps.

But if you place a human brain, with its evolutionary mandate to adapt to its environment, in an environment where there is no obvious linear sequence, where facts can be accessed at random, where everything is reversible, where the gap between stimulus and response is minimal, and above all where time is short, then the train of thought can be derailed. Add in the sensory distractions of an all-encompassing and vivid audiovisual universe encouraging a shorter attention span, and you might become, as it were, a computer yourself: a system responding efficiently and processing information very well, but devoid of deeper thought.

Thirty or so years ago, the term "climate change" meant little to most people; now it is understood by virtually everyone as an umbrella concept encompassing a wide variety of topics, including carbon sequestration, alternative energy sources, and water use, to cite just a few examples. Some feel that we're doomed, others that the different problems are exaggerated, and still others that science can help. Climate change is therefore not only global and unprecedented but also multifaceted and controversial. When we turn to the question of how future generations will think and feel, "Mind Change" can be an equally useful umbrella concept.

The argument underlying the notion of Mind Change goes like this. The human brain will adapt to whatever environment in which it is placed. The cyberworld of the twenty-first century is offering a new type of environment. Therefore, the brain could be changing in parallel, in correspondingly new ways. To the extent that we can begin to understand and anticipate these changes, positive or negative, we will be better able to navigate this new world. So let's probe further into how Mind Change, just like climate change, is not only *global,* as we've just seen, but also *unprecedented, controversial,* and *multifaceted.*

UNPRECEDENTED TIMES

Humans adapt. It is what we do better than any other species. Accordingly, our predecessors have always had to embrace a changing world where new inventions and technologies have, in turn, driven lifestyles, insights, tastes, and priorities. So why should this digital age be any different?

The automobile, for example, had vast, life-transforming effects. Using this kind of analogy, you could view digital devices as just the latest in a long line of innovations exciting and disturbing at first, then ultimately incorporated into our lives as the driver of some new development that will always be hard for some traditionalists to accept. Take the printing press, its introduction to Europe by Johannes Gutenberg around 1439 was undeniably a giant milestone in the progress of civilization. It democratized knowledge, and the reactionary forces of the status quo just didn't like it—a parallel, you might argue, to those who seem to be technological Luddites nowadays. Books began to disseminate insight to ever greater numbers of individuals, who then could, and did, foment social change, which led to personal advancement and universal education. Even fiction invariably raised issues about the human condition that enabled the reader to see the world through the eyes of others in other eras and locations, all the better to appreciate and shape one's own perspective and self-understanding; how could anything ever be more transformational?

Then there was electricity. Up to the end of the nineteenth century, nighttime would have brought uncontrollable darkness; the only redress for our ancestors would have been candlelight to fend off whatever unknown perils, real or supernatural, might be lurking just beyond that feeble, flickering pool of light. Our ancestors' experience of daily life would, for much of the time, be one of half-formed shapes, half-light, and a helpless inability to control their surroundings. Imagine the cataclysmic difference when eventually that dark and sinister world was flooded with electric light. What kind of new thinking and mindset might have occurred? Whatever it was, it was clearly a dramatic revision of reality to which our species adapted, and which thereby changed us.

Let's move to a more recent development: television. From the time of its invention around the middle of the twentieth century, the concern was that television would be bad for children's brains, that they would get "square eyes" and stop reading and playing outside. However, since television broadcasts occurred only during limited periods in the evening, and since there was at the time a dominant culture of outdoor games, reading, and collective family meals, the TV in fact complemented an existing lifestyle rather than disrupting it. In one sense, rather than being an early forerunner to the home computer, the TV was more like the Victorian piano in being a means of cohesive family activity and interaction.

This is no nostalgia for the golden days gone by. The middle years of the twentieth century were physically uncomfortable and tough, and turning back the clock, even if it were in some way possible, is not an attractive proposition: who in their right mind would ever opt for an unheated bedroom with uncooperative layers of thin scratchy blankets? But those *were* different times. There was only one TV set to a household, and that was if you were lucky; at first, usually only one home on a street might boast of such a marvel, attracting endless visitors to share in the wonderment. And even into the 1960s, watching TV had a communal feel.

Nothing could have been further from the twenty-first-century scenario of a family member rushing in from work or school to sit for hours in voluntary solitary confinement in front of the screen. One of the big differences between the earlier technologies and the current digital counterparts is quantitative, the *amount* of time the screen monopolizes our active and exclusive attention in a way that the book, the cinema, the radio, and even the TV never have. The futurologist Richard Watson certainly thinks that the *degree* to which digital technologies are dominating our lives makes the crucial difference: "We've always invented new things. We've always worried about new things and we've always moaned about younger generations. Surely most of [this] is conjecture mashed up with middle-aged technology angst? I think the answer to this is that it's a little different this time. [Screens] are becoming ubiquitous. They are becoming addictive. They are becoming prescribed."[1]

It's not so much the physical ubiquity of screens that might now differentiate the appearance of the average home from its predecessors, but an invisible feature, inconceivable a decade ago, whereby family members can be constantly connected beyond the household more intimately than with the immediate family members with whom they live in close proximity. Each individual adult and child now owns multiple digital devices that they use for entertainment, socialization, and information.[2]

There is a push and pull, respectively, *toward* the cyberspace offered by, say, the isolation of the mobile device and/or the multifunctional bedroom and *away* from the erstwhile epicenter of the family. In the past, bedrooms were places of punishment to which a child would be exiled for bad behavior—a far cry from the havens they are regarded as by many young people today. The warm kitchen or drawing room where the nuclear family sat together was the primary forum for interaction and information, and it provided a framework and a timetable for daily existence. Now the world of the screen in the bedroom, or anywhere else, has in many cases offered an alternative context for setting the pace, establishing standards and values, offering conversations, and providing entertainment, while the nuclear family eating a meal together is becoming less central in the midst of more complex societal trends of divorce and remarriage, as well as more variable and demanding work patterns.

Beyond the all-pervasiveness of digital technologies compared to inventions from previous eras, another difference is the shift from technology as a means to its being an end in and of itself. A car gets you from place to place; a fridge keeps your food fresh; a book can help you learn about the real world and the people in it. But digital technology has the potential to become the end rather than the means, a lifestyle all on its own. Even though many will use the Internet to read, play music, and learn as part of their lives in three dimensions, the digital world offers the possibility, even the temptation, of becoming a world unto itself. From socializing to shopping, working, learning, and having fun, everything we do every day can now be done very differently in an indefinable parallel space. For the first time ever, life in front of a computer screen is threatening to outcompete real life.[3]

You wake. The first thing you do is check your smartphone (62 percent of us), and in all probability you'll be checking your phone within the first fifteen minutes of consciousness (79 percent of us).[4] In 2013, 25 percent of U.S. smartphone users ages eighteen to forty-four could not recall a *single* occasion during which their smartphone was not within reach of them or in the same room. After waking, you grab a cup of coffee and a Danish while checking out emails that may have come in overnight as well as sending some yourself. Let's say your job enables you to work from home, as some 20 percent of American professionals do;[5] you'll then get down to business. While you have your tasks up in front of you, you will also have Twitter open to follow your favorite celebrity, along with your Facebook page to ensure that you don't miss out on any news. You'll also need to keep checking your social network sites, such as Instagram updates or Snapchat, and taking quick photos of what you're having for lunch (time has flown), all at the same time as being on the alert for good old-fashioned text messages. Exhausted by all this multitasking while working, you then relax by watching a YouTube video that attracted a large number of views, or you download the latest episode of a TV show. Next it's time to place your grocery order and have more serious retail therapy with some online shopping. In 2011, 71 percent of adult U.S. Internet users bought goods online,[6] and the following year a comparable number, 87 percent, of U.K. adults ages twenty-five to forty-four were shopping online.[7] By 2017, online sales are projected to account for 10 percent of all retail sales in the United States. Needing stimulation, excitement, and escapism after it hits home how much money you've just spent, you'll then immerse yourself in a thrilling videogame, just like some 58 percent of all Americans.[8] But now you feel a bit isolated and in need of some company. So you check out social networking, but this time looking more closely at online dating sites. U.S. Internet users spend 22.5 percent of their online time on social networking sites or blogs.[9] More than a third of couples who married between 2005 and 2012 in the United States reported meeting their spouse online, with about half of these meeting through online dating sites and the rest through other online sites such as social networking sites and virtual worlds.[10] The real, physical world and what we do in it may be becoming less and less relevant, as traditional constraints of time and space are fading. And as each of us adapts to an unprecedented new dimension, what sort of individual might eventually emerge?

For certain, someone who is less attuned to the outdoors. Since 1970, the radius of activity for a child, namely the amount of space surrounding the home in which the child freely wanders, has shrunk by an astonishing 90 percent.[11] And this restriction on play is unprecedented. In his book *A History of Children's Play and Play Environments,* Dr. Joe Frost traces the history of children's play from their early records in ancient Greece and Rome to the present time and concludes that "children in America have become less and less active, abandoning traditional outdoor play, work and other physical activity for sedentary, indoor virtual play, technology play or cyberplaygrounds, coupled with diets of junk food."[12] The consequences of play deprivation and the abandonment of outdoor play may well become fundamental issues in the welfare of children.

The content of a screen-based lifestyle is unprecedented not only in how it shapes thoughts and feelings but also because of the corollary effects of *not* exercising and *not* playing and learning outside. While an increasing number of digital aficionados may eventually opt for mobile technologies exclusively, for the time being an appreciable amount of time is still spent sitting down in front of a computer screen. In any case, if we're busy texting on our mobile phones or tweeting, even if we're out walking, we're still less likely to be taking more strenuous physical exercise than we may otherwise do. A clear corollary of a sedentary disposition is that we put on weight. Obesity stems from many factors, including the wrong kind and quantity of food, but also from reduced energy expenditure. It is hard to specify a particular order of events: whether a child who doesn't much like sports will be more attracted to the screen or a screen lifestyle has an allure that trumps climbing a tree is a chicken-and-egg scenario that is impossible to resolve here. Rather, we need to look at the whole digital lifestyle, both the increase in time spent in two dimensions and the simultaneous decrease in time spent in three.

For example, I recently received an email from a father of two young children in Australia that sums things up in a really arresting way:

Last weekend I had an eye-opening moment where the children had been lazing around the house, using and fighting over technology. When finally I was able to coerce them out for a short walk, we took bikes and I watched with delight the laughter and fun the kids had purely riding up and down this one particular steep-ish dogleg bend on this quiet country road. The enjoyment, laughter, and giggles from one's children are truly music to the ears of a parent. I do not ever hear that laughter when they are using technology.

A former teacher, Sue Palmer, flagged this issue back in 2007. Her book *Toxic Childhood* contained a list of simple activities that a child should have experienced before reaching adolescence, such as climbing a tree, rolling down a really big hill, skipping a stone, and running around in the rain.[13] How sad it is that these childhood activities, which would have been taken for granted a generation or so ago, should now be listed as identifiable goals that might otherwise not be achieved. Meanwhile, in a recent National Trust report, the term "nature deficit disorder" was coined not to describe a genuine medical condition but as a vivid expression of an endemic pattern of behavior, indicating for the first time ever that we have become dissociated from the natural world with all its beauty, complexity, and constant surprise.[14] Even the most diehard digital zealot cannot escape the simple fact that every hour spent in front of a screen, however wonderful, or even beneficial, is an hour spent *not* holding someone's hand or breathing in sea air. Perhaps even simply being at ease and happy in total silence could become a rarefied commodity that, instead of being a normal part of the human repertoire, will find itself on a wistful wish list of the future.

Professor Tanya Byron, a British psychologist best known for her work as a child therapist on television, was initially concerned specifically with regulation of the Internet; however, only two years later she recognized that the issue was not merely one of doing no harm but one of identifying the best possible environment beyond screen experiences. "The less children play outdoors, the less they learn to cope with the risks and challenges they will go on to face as adults," she wrote. "Nothing can replace what children gain from the freedom and independence of thought they have when trying new things out in the open."[15] In the past, play was most usually outside in fields and woods or in urban backstreets. Just look at the many books from the children's author Enid Blyton, written around the mid-twentieth century, where the young heroes and heroines were so busy catching smugglers and other shady villains that they only ever went indoors to have tea and to sleep.

At that time, in both fiction and fact, the environment in which you happened to be growing up provided a backdrop and props, not the actual narrative. The story came from inside your head—it had to—and arose from interaction with your friends as you became a cowboy or an Indian. It was the same inside the home, as plots were devised and story lines emerged from playing with dolls or toy soldiers or from dressing up. Trees, drawing pads, and toys (typically along with the cardboard boxes the toys came in) were merely tools and prompts for *your* game, *your* story, *your* internally driven scenario—above all, for *your* imagination. Sometimes, even quite regularly, you might be bored. But it was that very state of understimulation that impelled you to draw a picture, make up a game, or go outside to play. The point I want to stress is that *you* were the driver and *you* would be in control of your own inner world, your own private reality.

But now the screen can be the driver. Admittedly, you have to be mildly proactive in turning the device on and navigating your options, but once you have selected an activity, spectacular cyberexperiences contrived by someone else engulf you. You are now a passive recipient, and even though games such as The Sims, for example, allow you to modify and create worlds, it is always within the secondhand parameters of the game designer's thinking. I wonder how much of the time that previously would have been spent walking in the fresh air, playing the piano, or having a face-to-face conversation has now been forfeited in favor of a cyberactivity, a completely new type of environment where taste, smell, and touch are not stimulated, where we can be completely sedentary for long periods of time, yet where the ensuing experience trumps more traditional ways of life for appeal and excitement.

It would be simplistic in the extreme to think of the powerful and pervasive new digital lifestyle as either the apotheosis of human existence or the most toxic culture ever. We are being offered an unprecedented and complex cocktail of opportunity and threat, but not everyone is likely to agree on exactly what constitutes which.

A CONTROVERSIAL ISSUE

The American journalist H. L. Mencken once quipped, "For every complex problem there is an answer that is clear, simple, and wrong." Agonizing over whether digital technology is "good" or "bad" for the human mind is about as meaningless as arguing over whether a car is "good" or "bad." Nonetheless, debates on the complex issue of Mind Change are inevitable, as they will question the way we live our lives and the kind of people we might end up being. Rather than adopt simplistic and entrenched stances of "good" or "bad," "right" or "wrong," we need first to see where the various battle lines are actually being drawn, and then how we might resolve any resultant conflict in understanding and expectation.

Inevitably the biggest controversy revolves around the basic question of evidence: how strong it is and what it's actually demonstrating. Two reports in particular, surveying the evidence over the last few years, have suggested a "glass half-full" state of affairs. One was authored by psychologist Professor Tanya Byron in 2008 on the risks that children face from the Internet and videogames.[1] Her report came to the unsurprising conclusion that "the Internet and videogames are very popular with children and young people and offer a range of opportunities for fun, learning and development." However, Byron had concerns over potentially inappropriate material, ranging from violent content to the behavior of children in the digital world. She also drew attention to the notion that we shouldn't just be thinking about a child with a digital device in isolation, but should realize that the wider lifestyle picture is highly relevant, not least the child's relation to the parents.

The generational digital divide means that parents do not necessarily feel equipped to help their children in this unfamiliar space, which can lead to fear and a sense of helplessness. This sad state of affairs can be compounded by a wider risk-averse culture that is increasingly disposed to keep children indoors despite their developmental needs to socialize and take risks. While a risk-averse culture cannot by any means be the result exclusively of screen living, it obviously provides an attractive incentive and alternative for a child to be readily persuaded not to venture outside. Another uncontroversial point made by Byron's report was that while children are confident with the technology, they are still developing critical evaluation skills and need adult help to make wise decisions. In relation to the Internet we need "a shared culture of responsibility."

Byron's real emphasis has been on protection, but her report also touched on the wider issue of the empowerment of children: "Children will be children pushing boundaries and taking risks. At a public swimming pool we have gates, put up signs, have lifeguards and shallow ends, but we also teach children how to swim." All that said, for the time being, anyone reading Byron's report would feel that there was no immediate need just now for any revolutionary, or even merely interceptive, action.

It was a similar story a little later in 2011, when neuroscientist Dr. Paul Howard-Jones of Bristol University was commissioned to produce a review on the impact of digital technologies on human well-being. Howard-Jones accordingly set about discussing what the field of neuroscience has established regarding the effects of interactive technologies on behavior, the brain, and attitudes, with a special focus on children and adolescents. After all, "the vanguard of our advance into this new world is our children, and especially our teenagers. We know that the developing brain of a child is more plastic, and responds more malleably to experience than an adult's brain."[2]

Commendably, Howard-Jones highlighted the need to understand the uses of technologies in a specific context rather than to label particular technologies, or technology more generally, with a blanket description of "good" or "bad." He also highlighted the findings that some technology-based training can improve working memory or provide mental stimulation that slows cognitive decline, while some types of gaming can improve visual processing and motor response skills. However, his review also identified three potential risks for children: violent videogames, the use of games and other technology leading to sleep problems, and excessive use of technology having a negative physical or mental impact or interfering with daily life. He went on to point out that any changes in the mindset of the upcoming generations are, most crucially, anticipating changes in society as a whole—so the issues are relevant to all of us, whatever our age.

These snapshots from Byron and Howard-Jones depict an image of the Digital Native that is still currently blurred and uncertain, yet cautiously sanguine. Both reports leave at best an overall feeling of reserved optimism and at worst the usual academic-type conclusion that the jury is still out because "more research is needed." Both Byron and Howard-Jones paint an equivocal but generally positive picture of work in progress, so long as we are constantly alert to ever-present dangers such as bullying, sexual grooming, and violent gaming. Any concerns either author has have more to do with regulation. On the whole, the conclusions in both cases err on the side of the mildly positive with regard to learning, socializing, and improving mental function. The glass is half full, so long as everyone acts sensibly.

But such comforting assessments seem significantly outnumbered by voices from various professionals around the world who were not commissioned to provide a generalized snapshot of the current moment but instead deal with what happens when the use of digital technologies is *not* sensible. The glass then appears half empty.

First, there's the perspective articulated in books such as *iDisorder* by clinician Larry Rosen[3] or *Alone Together* by MIT psychologist Sherry Turkle,[4] who suggest that the more people are connected online, the more isolated they feel. In both cases, the concern is for when Internet use becomes obsessive. Perhaps surprisingly, captains of the digital industries themselves are also worried. Biz Stone, the cofounder of Twitter, made headline news by stating at a conference: "I like the kind of engagement where you go to the website and you leave because you've found what you are looking for or you found something very interesting and you learned something."[5] The idea would be that you use Twitter to enhance the quality of your real life. But even he believes that using Twitter for hours at a time "sounds unhealthy," presumably because it means his invention has become a lifestyle in itself. Then there's Eric Schmidt, erstwhile CEO and now chair of Google: "I worry that the level of interrupt, the sort of overwhelming rapidity of information . . . is in fact affecting cognition. It is affecting deeper thinking. I still believe that sitting down and reading a book is the best way to really learn something. And I worry that we're losing that."[6]

This worry is prescient in the light of what many neuroscientific and medical experts are voicing.[7] For example, neuroscientist Michael Merzenich, one of the pioneers in demonstrating the incredible adaptability of the nervous system, has concluded, in the typically restrained language required of his profession: "There is thus a massive and unprecedented difference in how their [Digital

Natives'] brains are plastically engaged in life compared with those of average individuals from earlier generations, and there is little question that the operational characteristics of the average modern brain substantially differ."[8]

Educators are also voicing worries. In a 2012 report that surveyed four hundred British teachers, three-quarters reported a significant decline in their young students' attention spans.[9] In the same year, a survey of more than two thousand U.S. secondary school teachers showed that 87 percent of teachers believed that digital technologies are creating an "easily distracted generation with short attention spans," whereas 64 percent agreed that these technologies have more of a distracting effect than a beneficial one on students academically.[10] The diversity of different professions expressing the drawbacks of digital devices was well illustrated in an open letter written in September 2011 to the respected British newspaper the *Daily Telegraph* and signed by two hundred teachers, psychiatrists, neuroscientists, and other experts expressing alarm over the "erosion of childhood."[11]

However, perhaps one of the most telling surveys has been to target aficionados of cyberspace themselves. The Pew Research Center in the United States, along with Elon University, asked more than one thousand technology experts how the brains of "millennials" (a term pretty much interchangeable with "Digital Natives") will change by 2020 as a result of being so connected to online digital technologies.[12] These professionals were asked which of two predictions was the more likely for the immediate future, as articulated in two contrasting statements. One was extremely positive:

> Millennials in 2020 do not suffer notable cognitive shortcomings as they multitask and cycle quickly through personal- and work-related tasks. They learn more and are adept at finding answers to deep questions, in part because they can search effectively and access collective intelligence

via the Internet. Changes in learning behavior and cognition generally produce positive outcomes.

The other was more negative:

Millennials in 2020 do not retain information; they spend most of their energy sharing short social messages, being entertained, and being distracted away from deep engagement with people and knowledge. They lack deep-thinking capabilities; they lack face-to-face social skills; they depend in unhealthy ways on the Internet and mobile devices to function.

The group of digital experts was split rather evenly on what they predicted for the future. But perhaps most tellingly, many of those who went along with the positive prediction noted that it was *more their hope than their best guess.* So even the 50 percent or so of professionals who regard the screen culture in a favorable light overall do so, in many cases, from a stance of wishful thinking rather than of certainty or rational argument.

Further evidence indicating that something might be going awry is perhaps every bit as compelling as expert opinion or epidemiological and experimental research: the very apps and websites that point to clear trends in the tastes and proclivities of current society. One app, paradoxically called Freedom, will block your Internet access for a user-specified amount of time each hour, while SelfControl will enable you to bar yourself from websites that you feel you are following too slavishly but are helpless to resist. Zadie Smith, author of the acclaimed bestseller *White Teeth,* for instance, credits these two Internet applications in the acknowledgments section of her latest work.[13] Apparently she was struggling to maintain her concentration while writing her new book because of the diversions available just a click away on the Internet. So she was grateful to the apps for "creating the time" in which she could write.

And Zadie Smith is not alone. The success of these flourishing enterprises obviously raises the question of why they are doing so well. Why should increasing numbers of people require some external service to stop them from using the Internet, rather than just switching it off for themselves? As with junk food or cigarettes, we become addicted to the distraction of an external input that determines and shapes our actions, choices, and thoughts. The existence of these apps in themselves does not mean that there's an epidemic of screen addiction, but it *does* imply that there are enough customers who experience these problems for the apps to be profit-making enterprises. We cannot ignore that even the platforms and users themselves implicitly acknowledge that screen technologies can be something we use compulsively.

Another unprecedented feature of our current society is the lightning-speed dissemination of information. The hyperconnected blogosphere reaches more people more quickly than satellite radio and television: the Pakistani citizen who unwittingly tweeted live updates of the raid on Osama bin Laden's house was able to access a large audience more quickly than any other form of media. Yet, for precisely that reason, the blogosphere is the perfect medium for spreading misinformation relating to complex issues, or even for just oversimplifying them. Such is the concern of the World Economic Forum's Risk Response Network, which provides leaders from the private and public sectors with an independent platform to map, monitor, and mitigate global risks. Its 2013 annual *Global Risks Report* analyzed the perceived impact and likelihood of fifty prevalent global risks over a ten-year time horizon; among those listed is "digital wildfires in a hyperconnected world."[14]

I first joined the fray over the impact of digital technologies back in February 2009 with my speech in the House of Lords (described in the preface to this book) on the possible unexpected effects on the human mind of social networking.[15] All I did was make the neuroscientific case for the well-accepted plasticity of the brain and point out that new types of screen experience would likely have a new type of impact on mental processes. The reaction, worldwide, was disproportionate to the tentative syllogism I was putting forward. While some seemed to agree with me, others were emphatic in insisting that there was "no evidence" for what I was saying.

While one might think this issue of evidence would be an easy matter to resolve, the problem with a simple negative argument is that even if there were no scientific findings at all to back it up, absence of evidence is not evidence of absence. In science, you can only conclusively establish with experiments that a finding is positively the case, never the reverse. After all, it might simply be that the test you are using isn't the most appropriate, or that the measuring instruments are not sensitive enough, or that the effects will be delayed or too immediate to fit your particular observation period. The point is that you cannot be conclusive, and you must therefore leave open the possibility that there is indeed an effect, albeit one that you haven't been able to detect. Thus it is impossible to demonstrate definitively that screen-based activities have no effect at all on the brain or behavior, any more than I or anyone else could prove definitively, to use an age-old example, that there is *not* a teapot in orbit around Mars.

This constraint poses a problem for both sides, since it is impossible to demonstrate just as conclusively that screen-based activities *are* having an unequivocal effect on the brain and consequent behavior.[16] Let's assume a finding is reported of some definite effect, good or bad. Even then, in the evaluation of scientific findings, few single peer-reviewed papers, that gold standard of professional probity, are viewed unanimously by all scientists as conclusive. It is normal practice for research to continue, and for interpretations to be revised as results accumulate. Interpretations of the evidence are inevitably subjective, with different scientists placing different emphases on different aspects or priorities within the experimental protocol. There is very rarely a Rubicon that, once crossed, means that a finding is universally accepted as the "truth." Truth is always provisional in science, waiting for the next discovery to come along that could displace the current view (or, as it would by then be disparagingly called, "the current dogma"). When enough doubt accumulates to challenge this dogma, when accepted patterns of thought are straining to account for just too many anomalies, the reappraisal of what is true amounts to a "paradigm shift"—a concept Thomas Kuhn first introduced in 1962 in his now classic work *The Structure of Scientific Revolutions.*[17]

A wonderful example of how scientists can stick rigidly to dogma and have closed minds to highly novel ideas is the revolution in treatment of ulcers that developed in the 1990s. The hero of the story is an Australian physician, Barry Marshall. As part of his training, Marshall was working in a lab with another scientist, Robin Warren, studying bacteria. Contrary to accepted dogma, they found that a certain bacteria, *Helicobacter pylori,* could survive in a highly acidic environment, such as the stomach. Marshall and Warren started to doubt the well-accepted and established body of knowledge that ulcers were caused by excessive acid and thus were primarily the result of stress. What if ulcers were the result of bacterial infection instead? What would happen to the blockbuster drugs currently on the market for ulcers but perhaps designed for the wrong biological target? The implications for the pharmaceutical industry, as well as for the medical establishment, were huge. "Everyone was against me," Marshall recalls.[18] For many years, good old unscientific prejudice delayed significantly the final acceptance of Marshall and Warren's theory. Starved of funding but convinced of the merits of their theory, Marshall actually drank a glass of the medium containing the bacteria and duly gave himself an ulcer, which was cured by antibiotics. Vindicated at last, he and Warren won a Nobel Prize.

Even without the need to wait for a seismic paradigm shift, disagreement is fundamental to science: what one individual re-searcher will see as an exciting discovery, another may view as an epiphenomenon, while a cynic might regard it as unproven. It is not in the act of empirical observation but in the consequent sub-jective evaluation that there is most room for controversy and doubt. In all branches of science, the explanation that is formu-lated as scientists pore over the latest data is never conclusive. Any scientist writing the discussion section to wrap up a paper for a peer-reviewed journal will invariably be tentative and provisional, always remembering that not all potentially salient facts and fac-tors are known. Scientists inhabit a hesitant world that is far from absolute, where doubt is as natural as breathing. So while disagree-ment in science is normal and unavoidable (if not necessarily un-derstandable at first), the flat refusal even to debate and to think about possibilities, as can happen with the question of screen tech-nologies, is not.[19] The only realistic way forward is to plow through as many individual papers as possible that each tackle a specific issue and that collectively form a general overall picture.

In the case of cyber-induced long-term changes in the brain and resultant behavior, we are faced with a complex situation, one not amenable to a definitive litmus test or a single smoking-gun experi-ment. What kind of evidence might one hope for, in a realistic pe-riod of time, that could demonstrate to everyone's satisfaction that screen culture is inducing long-term transformations in wide-ranging phenomena as diverse as empathy, insight, understand-ing, identity, and risk taking? What single, one-off finding would it take for those who resist the possibility that there just might be something amiss after all, or at least that we are missing opportuni-ties?

Concepts such as Mind Change are, in Kuhn's terminology, par-adigms, not specific single hypotheses that can be empirically tested in highly constrained and specific experiments. An umbrella concept such as Mind Change, as we're about to see, draws together threads from apparent societal trends and expert professional views, as well as a wide range of direct and indirect scientific findings from different disciplines. The majority of the scientific studies reported in the chapters to come have been peer reviewed; this process ensures that they have demonstrated "statistically significant" findings, which means they are not subjective judgments but the results of a standardized and well-established system of testing.[20]

Irrespective of the different types of evidence that support it, inevitably the notion of Mind Change as a new paradigm has stirred up allegations of scaremongering and inciting moral panic. But bear in mind that scaremongering is predicated on the notion that there is really nothing to be scared of in the first place. Do we in fact know that this is the case? However, if and when the validity of the scare is irrefutably demonstrated, then the scare turns into an established danger. So now the original prediction would actually have been something very different, a wake-up call. Dismissal on the grounds of scaremongering should be, if anything, a final conclusion and not an opening gambit.

As for moral panic, perhaps any criticism of the digital world could be interpreted by aficionados of cyberspace as an attack on their personal lifestyle and therefore ultimately on them as individuals. But there is no need to panic at the moment. Indeed, if we allow ourselves the opportunity to take stock of where we are and where we wish to go in the twenty-first century, we can work out what our lifestyle and society need to look like in order to get us there. But to do that we first need to unpack the various very different issues that Mind Change embraces.

A MULTIFACETED
PHENOMENON

Climate change, according to the Intergovernmental Panel on Climate Change, "may be due to natural internal processes or external forcings, or to persistent anthropogenic changes in the composition of the atmosphere or in land use."[1] Nobody could dispute that there's a multitude of issues involved here. So it is with Mind Change, which I'm suggesting is comparably multifaceted, throwing up a range of different questions that need to be explored independently. These different questions fall into three main areas, worth previewing here: social networking and the implications for identity and relationships; gaming and the implications for attention, addiction, and aggression; and search engines and the implications for learning and memory.

In no particular order of priority, let's start with social networking. A recent radio program on the BBC featured Kaylan, an eighteen-year-old man who decided to take advantage of the opportunity offered by Facebook as of September 2011 to remove all privacy settings on his page so that any number of followers could track his daily life in the public domain. He boasted of having some one hundred thousand followers at the time of the broadcast. Kaylan also admitted he had done nothing at all to deserve fame. His posts were often mundane photos of himself throughout the day leading a "crazy life."

So what was it that was so attractive to his followers? Well, there were a whole bunch of similar folk like him who could engage in arguments with each other. Then the followers could take sides. Yes, Kaylan had his fair share of "haters." After all, he added, "you can't be nice on Facebook." By saying unpleasant things such as "Kill yourself," these haters would then garner maximum praise and "fame" for themselves. While Kaylan is obviously very far from being a typical Facebook user, both he and his hundred thousand followers serve as an example of the unprecedented extremes to which the medium can be taken. Your importance as revealed by social networking activity can even now be quantified.[2]

The majority of Facebook users are far less dramatic. Still, in a Pew Research Center survey, U.S. social network users aged twelve to nineteen overwhelmingly chose negative rather than positive adjectives to describe how people act on social networking sites, including "rude, fake, crude, over-dramatic and disrespectful."[3] For example, one middle school girl respondent commented: "I think people get, like when they get on Facebook, they get ruthless, stuff like that. . . . They act different in school and stuff like that, but when they get online they're like a totally different person, you get a lot of confidence." Another girl said: "That's what a lot of people do. Like they won't say it to your face, but they will write it online."

A recent meta-analysis looking at data collected over thirty years from fourteen thousand U.S. college students indicates that overall levels of empathy may be declining, with an especially steep drop in the last ten years—a time frame that corresponds well with the advent of social networking among Digital Natives.[4] Of course, a correlation is not a causal link, but this is just the type of close correspondence that should serve as a starting point for rigorous epidemiology to establish whether there might be a direct causal link between screen time and a reduction in empathy. We should also be asking why those who already have problems empathizing, such as individuals with autistic spectrum disorder, are particularly comfortable in the cyberworld. More generally, could this sanitized and limited type of interaction account for the ease with which bullying, always a dark part of human nature, has now found unconstrained expression in the cyberworld? After all, if you haven't rehearsed the basic nonverbal communication skills of eye contact, voice modulation, body language perception, and physical contact, you won't be particularly good at them, and it will be harder for you to empathize with others.

More than a billion people worldwide use Facebook to keep in touch with friends, share pictures and video clips, and post regular updates of their movements and thoughts.[5] Another estimate is 12 percent of the entire global population, with 50 percent of North Americans, 38 percent of antipodeans, 29 percent of Europeans, and 28 percent of Latin Americans signed up.[6] (These figures are based on total population; if we exclude newborn babies, the severely infirm, and others with no computer access, the number of Facebook users as a proportion of the computer-using population is probably far higher). A further two hundred million actively use Twitter, the "microblogging" service that lets users circulate short messages about themselves, post pictures, and follow the minutiae of others' stream of consciousness or daily routines.[7]

Nowadays, all generations are represented on the sites, with octogenarians able to stay in touch with grandchildren living far away, but it is the Digital Natives who are the most avid users. In the United Kingdom, 64 percent of adult Internet users age sixteen and over are social network site users, whereas 92 percent of those aged sixteen to twenty-four who use the Internet have profiled themselves on a social networking site.[8] In the United States, 80 percent of online teens ages twelve to seventeen use social networking sites, mostly Facebook and MySpace.[9] U.S. users have on average 262 friends,[10] a figure higher than the world average of roughly 140 friends.[11] Twelve- to twenty-four-year-old Facebook users have, on average, more than five hundred Facebook friends.[12] Roughly 22 percent are from high school, 12 percent are immediate family, 10 percent are co-workers, 9 percent are from college, and 10 percent of friends have never been met in person or only been met once.[13]

On an average day, 26 percent of Facebook users "like" a friend's status, 22 percent comment on a friend's status, while only 15 percent update their own status.[14] So more people spend time interacting with other users' content rather than posting their own. All of which points to a blindingly obvious truth: social networking has become a central factor in the culture of all but the very poorest and most deprived regions in the world, or the most ideologically repressed. A critical question then is, quite simply, what is so special about social networking? What is the basic need that this new culture is meeting in an apparently unprecedented and yet effective way? If we are to understand and appreciate the changing mind of the mid-twenty-first century, this is one of the most important questions to ask.

The benefits of social networking seem irrefutable: direct marketing to the consumer, dating sites, career building, contact with old friends. Being "connected" is often cited with an enthusiasm that automatically assumes it is a desirable scenario. But what worries me is whether this almost incessant communication through the screen might have a downside as well. As always, there's the key issue of being "sensible": while social networking sites could provide harmless fun and complement real friendships if they are used in moderation, if they are used excessively or to the exclusion of real relationships, they might perhaps impact in a very fundamental and unforeseen way on how you view your friends, friendship, and ultimately yourself.

If you're increasingly anchored in the present and consequently devoting all your time to the demands of the outside world, a robust sense of inner identity might be harder to sustain. Perhaps the constant accessing of social networking sites will mean living a life where the mere thrill of reporting and receiving information completely trumps the ongoing experience itself—a life where checking in at a restaurant, posting pictures of a meal, and yearning for "likes" and "comments" generates more excitement than the occasion of dining out itself. The momentary exhilaration you'd be feeling would shift from being generated by a firsthand life experience toward the slightly delayed indirect experience of the continuing reaction and approval of everyone else. If we're going to be living in a world where face-to-face interaction is less practiced and is thereby uncomfortable, then the "push" of such an aversion to messy real-life, three-dimensional communication, combined with the "pull" of the appeal of a more collective identity of external reassurance and approval, may be transforming the very nature of personal relationships. The knee-jerk speed required for reaction and the reduced time for reflection might mean that those reactions and evaluations themselves are becoming increasingly superficial: already people are using phrases such as "kill yourself" and "hater" on Facebook in a context that conveys far less depth of real feeling and of individual background history than these terms previously would have implied.

Privacy appears to be becoming a less prized commodity: among the U.S. young aged thirteen to seventeen, more than half have given out personal information to someone they don't know, including photos and physical descriptions.[15] Meanwhile, Digital Natives post personal information on their Facebook page that is typically shared with more than five hundred "friends" at a time, fully aware that each of these friends could then pass on that information to hundreds more in *their* networks.

It has become more important to have attention, to be "famous." The trade-off for such fame is, and always has been, as the mid-twentieth-century film star Greta Garbo famously exemplified in her repeated pleas of "I want to be alone,": loss of privacy. So why is it that the privacy we so treasured previously we now hold in increasingly casual disregard? Until now, privacy has been the other side of the coin to our identity. We have seen ourselves as individual entities, in contact with the outside world, yet distinct from it. We interact with that outside world, but only in the ways and at the times we choose. We have secrets, memories, and hopes to which no one else has automatic access. This secret life is our identity, distinct from a professional one and even more intimate than a private life of individual friendships in which we vary what and how much we confide to others. It is a kind of inner narrative that, until now, has provided each individual with his or her own way of linking past, present, and future—an ongoing subjective, internal commentary that meshes past memories and future hopes with the happenstance of each day. Now, for the first time, this secret story line is being opened up to the outside world, to an external audience that can be uncaringly capricious and judgmental in its reaction. A particular identity therefore no longer is an internal, subjective experience, but is constructed externally and therefore is much less robust and more volatile, as has already been suggested in a recent report to the British government on "future identities."[16]

A second cornerstone of the digital lifestyle is gaming. In the mid-1980s, children might have spent about four hours a week on average playing videogames at home and in arcades.[17] But fast-forward a decade or so and the videogame has become an integral part of the home scene and beyond.[18]

A 2012 study of U.S. adolescents reported that boys between the ages of ten and thirteen were playing on average a staggering forty-three hours a week (although, admittedly, the number of subjects was fairly small, 184).[19] Yet even conservative estimates (from 2009) indicate that the average U.S. child between the ages of eight and eighteen is spending seventy-three minutes a day recreationally in this one screen-based activity, up from twenty-three minutes in 1999.[20] That means at least an hour a day spent not interacting with the real world, and in particular not studying. In a survey of U.S. youth between ages ten and nineteen, gamers spent 30 percent less time reading and 34 percent less time doing homework.[21] Granted, it is hard to separate the chicken from the egg: perhaps children who perform more poorly in school are likely to spend more time playing games, which may give them a sense of mastery that eludes them in the classroom. We need to go beyond correlation to cause, but what we can't do is just ignore the issue altogether.

Videogames open up fertile territory for controversy. On one hand, there are clear positives, as we'll explore in detail later: for example, improved sensorimotor coordination and perceptual learning. On the other hand, various and many stories around the world can paint a terrible picture of a modern lifestyle of overindulgence in the unfettered fun of playing videogames. For example, in Taiwan in February 2012, a twenty-three-year-old man was found dead in an Internet cafe after twenty-three hours of continuous gaming.[22] Another young man in Taiwan, age eighteen, died in July 2012 after forty hours of continuous gaming.[23] Then there was the report of two parents neglecting their own real baby, who subsequently died, in order to raise an online virtual baby.[24] In December 2010 a man in the north of England received a life sentence after killing a toddler immediately after losing in a violent videogame.[25] Then there was the case of a gamer who hunted down his virtual opponent in real life and stabbed him as revenge for being stabbed in the game.[26] And this is not to mention the list of high-profile suicides of gamers.

The immediate defense of a gaming fan would probably be that (1) this is all scaremongering and unlikely to be true; (2) it is unlikely to be the whole story, with other, more important factors really to blame or to mitigate the circumstances; or (3) these examples, horrendous though they may be, are isolated cases that are actually extremely rare. All of these possibilities are not mutually exclusive and may indeed be the case, but they should be conclusions, not starting premises. Moreover, even if such stories are exaggerated and uncommon, they may still be important as caricatures of certain prevailing trends now emanating from society, albeit in a much milder form: a profile of addiction, aggression, impulsivity, and recklessness.

Modern gamers enter a visually rich world where they can assume a character completely unlike themselves or, in some games, create whatever kind of character (avatar) they desire. They navigate these fictional beings through situations involving moral choices, violence or aggression, and role playing, with intricate reward systems built into the games that provide the incentive to carry on living out the fantasy. Some individuals can become so immersed that they lose track of the real world and time; they report that they turn into their avatars when they load the game. Alternatively, gamers may develop an emotional attachment to their character. So how are these highly stimulating, often violent games with possible addictive qualities actually affecting us?

One outcome could be enhanced aggression. Experimental studies are revealing that violent videogames lead to increases in aggressive behavior and aggressive thinking accompanied by decreases in prosocial behavior.[27] It seems that videogame-induced aggression is directly caused not only by immediate provocation but also by more indirect biological predispositions and environmental influences, as an individual gradually develops a more adversarial worldview. Although violent games have not been proved to be the immediate trigger for criminally violent behavior, there is strong evidence that playing them may increase the type of low-grade hostility that occurs every day in schools or offices.

It may also be that videogames lead to excessive recklessness. In one recent investigation using brain imaging, the key finding was of an enlargement of a specific area of the brain (the nucleus accumbens) typically seen in the brains of compulsive gamblers.[28] Most intriguing of all is that this particular brain region releases dopamine, a key chemical messenger whose production is increased by all addictive psychoactive drugs. These chemical similarities between the brains of gamers and those of gamblers do not prove that gaming is technically addictive, but both may well share a further feature: recklessness. After all, it is a dangerous lesson to learn that death lasts only until the next round—it may suggest that actions in the real world don't have real consequences.

The crucial factor once again will be whether an individual is being, in the minister's words in our House of Lords debate back in 2011, "sensible and proportionate" about playing games. It's a bit like eating chocolate: the occasional treat in an otherwise balanced diet is relatively harmless and enjoyable, whereas an unremitting daily diet exclusively of chocolate would have dire consequences. The problem is not with those who might play games occasionally as a pastime in a portfolio of other interests and activities in the real world, but the number of frequent gamers who, from the amount of time they spend on gaming to the exclusion of all else, end up obsessional or hooked.

Finally, in addition to social networking and gaming, there's a third aspect to Mind Change: surfing the Internet, particularly with search engines. If you are not using digital technologies interactively to engage in a relationship or to play a game, then the screen can still have intoxicating appeal simply because of what it can tell and show you—some might go so far as to say teach you. It's almost unbelievable that this essential facility started less than twenty years ago, in 1994, when Yahoo! was created by Stanford University students Jerry Wang and David Filo in a campus trailer, originally as an Internet bookmark list and directory of interesting sites. Then in 1996 Sergey Brin and Larry Page, two Stanford University students, tested Backrub, a new search engine that ranked sites according to relevance and popularity. Backrub was destined to become Google, which currently has around 80 percent of the global market share in search, while its nearest competitors are in single digits.[29] The brand name has become a verb: almost everyone "Googles."

Sometimes, for no obvious reason, seemingly pointless activities, striking a funny posture, "planking" or performing a little dance such as the Harlem Shake, draw crowds. I have my own direct experience of how powerful such viral phenomena can be. Back in April 2010 I was being interviewed by Alice Thomson of the UK *Times* about the impact of digital technology on how we feel and think. We had progressed to discussing how fast-paced technology might mandate correspondingly fast views and reactions. Trying to provide her with a sound-bite summary, I raised the prospect of humans being reduced to simple negative or positive gut reactions, such as "yuck" or "wow," to whatever flashed on the screen. Because I tend to talk quickly, Alice misheard and transcribed what I'd said as "yaka-wow." This may be amusing enough in itself, but the point is that just twenty-four hours later one could find seventy-five thousand results for this term on Google. Moreover, someone bought the domain name, and soon I was astonished to see mugs and T-shirts sporting the term "yaka-wow." On one website, the First Church of the Yaka-Wow welcomed "breezy people to a world of no consequences." The term had gone viral within a time frame that would have been unthinkable only a decade or so earlier.

So what is the potential of digital technologies to help everyone, of any age, to learn things, in the broadest sense of the term? Presumably, when people surf they are feeding into a search engine specific terms or names, if not formal questions, and receiving relevant information in response. They are "learning." The dictionary defines learning as "the act or process of acquiring knowledge or skill." The current digital technology may enhance this ancient, superlative human talent, or then again it may jeopardize it, but we need to unpack the various issues involved. The appeal of the surfing experience, the differences between silicon and paper, the educational value of digital technologies, and, above all, access to a nearly infinite amount of information all operate as different and unprecedented factors to shape our thought processes.

Search engines are now part of our lives, and for many they are the immediate and obvious first stop for finding out a fact or learning more about a subject. So screens could shape our cognitive skills in a fundamentally new way. Surely one of the most important issues to explore is whether the next generation might be learning very differently compared to their predecessors who used books. The most obvious difference is a tactile one—we handle paper much more differently than we do screens. That being so, how might the pleasures of reading on a screen match up to those of paper? Flicking pages back and forth, highlighting sentences, and scribbling in the margin may all be positive features that contribute to the absorption of what you are reading, so the potential for personal interaction with a paper book may be greater than with a screen.

Anne Mangen at the University of Oslo explored the importance of actually touching paper by comparing the performance for readers of paper compared to readers of the screen. Her investigation indicated that reading on a computer screen entails different strategies, covering everything from browsing to simple word detection, that together lead to poorer reading comprehension in contrast to reading the same texts on paper.[30] Moreover, apart from the physical features of the printed page compared to the pixelated one, the screen can have an additional feature that the book can never have: hypertext. Above all, a hypertext connection is not one that you have made yourself, and it will not necessarily have a place in your own unique conceptual framework. Therefore, it will not necessarily help you understand and digest what you're reading, and it may even distract you.

But the whole point of screens is not simply that they can serve as substitute books. A still deeper issue is how computers, tablets, and e-readers can provide information in an utterly different, nonverbal way, and thereby perhaps actually transform how we think. If inputs arrive in the brain as images and pictures rather than as words, might that, by default, predispose the recipient to view things more literally rather than in abstract terms?

These, then, are the ever more invasive and pervasive technologies that have the power to transform not just what we think, but how. Yet Mind Change involves more than innovative gadgetry: just as critical is the mind that is to be changed. It is the growth and connections between the brain cells we are born with that will turn us into the unique beings we are, with brains capable of individual and original thought. There are many talents we as a species lack: we don't run particularly fast or see particularly well, nor are we particularly strong compared to others in the animal kingdom. But our brains have the superlative talent to adapt to any environment into which we are placed, a process known as plasticity. As we make our personal, idiosyncratic way through life, we develop our own particular perspective as a consequence of these personalized connections in our brains. It's this unique pattern of connectivity that I'd like to suggest amounts to an individual mind. So in order to appreciate the impact of these global, unprecedented, controversial, and multifaceted technologies on the twenty-first-century human mind, we need next to look through the prism of neuroscience.

HOW THE BRAIN WORKS

How could an experience, screen-based or otherwise, literally leave its mark on a sludgy brain? If we neuroscientists are to contribute anything significant at all to appreciating the effects of the digital lifestyle on our mental processing, it's by pointing out the actual physical neuronal mechanisms at work: we should be able to demonstrate the causal link between exposure to certain environments and experiences, and ensuing thoughts and behavior. By understanding as much as possible about how the brain works, we'll be able to get a much more accurate picture of how and to what extent screen technologies could be transformational.

The big challenge for neuroscience has always been to make the intellectual leap between a bit of brain tissue and a thought, an emotion—even a dream, in both senses of that word: the literal phenomenon of that bizarre inner world that unfolds during sleep, as well as the metaphor for planning wonderful outcomes for our lives. It's a journey we'll need to make in three steps: first, to find out how the brain itself works; second, to discover how it changes throughout life; and third, to see how these changes in the brain could amount to the "mind." Yet it's far from obvious even where to start.

"So how *does* the brain work, then?" The girl in front of me, probably about eleven years old, was insistent. Surely it was simply because I had run out of time in my one-hour talk to her group of schoolchildren that I had omitted to clear up this final, trivial question. We had looked at the brain from all angles by taking apart a plastic model. I had told my young audience about the time when I had been a student myself and had held a real human brain in my hands and, because brain tissue is nothing like the hard, bright pink plastic model but is creamy white, soft, and fragile, I had pondered what would have happened if some of it got caught under my fingernail. Can a memory or an emotion be dislodged by a fingernail? Could a bit of brain tissue that somehow related to a particular habit, such as biting your fingernails, actually end up adrift *under* a fingernail? How is the experience of being you, of seeing the world in a way no one else can share firsthand, generated by this unappealing and uncooperative mass that you can cup in one hand?

No model brain, nor indeed its real-life counterpart, offers any obvious starting point. There are no conspicuous moving parts, as there are for the heart or the lungs, that indicate what is going on. All you can do by looking at the brain is appreciate how, on the macro level, it is put together. You'll see that there are enveloping layers around the top of the spinal cord as it swells out into the most basic part of the brain.[1] From there evolution has added further compartments and easily discernible structures—brain regions that vary in size and importance according to the species. But the theme is the same for all mammals, whether you're looking at the brain of a rat or of a human. You'll always see, for example, a small cauliflower-like growth coming out from the back of the brain just above the spinal cord.[2] You'll also always see the two hemispheres that jam against each other like two fists, with their outer covering, the cortex (Latin for "bark") wrapping around them the way bark wraps around a tree.[3]

The surface area of the cortex has expanded in humans to such an extent that accommodating such vast amounts of brain in the confines of the skull would be like accommodating a sheet of paper in a tight fist: you would have to crumple the paper up. In a sense, and so long as we don't stretch the analogy too far, this is what evolution has done: the surface of the human brain is as wrinkled as a walnut, that of the other primates less so, that of cats and dogs even less still, and the cortex of rodents not at all. This thin outer layer is perhaps the most fascinating and enigmatic part of the brain. In evolutionary terms, it is the newest and, perhaps not surprisingly, the most prominent in humans, the species with the greatest intellectual capacity. So the cortex will feature more than any other brain area as we explore the impact on thinking of the digital technologies.

To get an idea of how the brain is put together, think of a busy metropolis such as New York City. The anatomically distinct brain regions would correspond to boroughs, within which would be districts and then neighborhoods—in brain terms, smaller and smaller groups of cells. By the time we arrive at a block, a street, or a line of houses, we are at the basic unit of neuronal communication: the gap (synapse) between any one brain cell and another. And the house on the street? That would be the neuron itself, the rooms within it the organelles, the specialized cellular parts that keep a single brain cell alive, just like any generic cell in the body. While this metaphor may convey the nested hierarchy of the anatomy of the brain areas, the extrapolation can go no further: it is simply a static snapshot of how the physical brain is built up.

In my talk to the young students, I had pried the plastic model apart and shown them all the different and easily discernible regions beneath, how they intertwined around each other, just as I had first seen in a real brain so long ago in the dissecting room of the Oxford University Anatomy Department. But would that answer satisfy the little girl standing in front of me, eyes like saucers, impatient for me to tell her in a sentence how the brain worked? The problem is that brain cells are less analogous to fixed structures such as bricks and houses, which don't actually do anything, and more comparable to people, their highly dynamic inhabitants. What we really need, therefore, is an image, some kind of scenario that describes not only how the brain is constructed anatomically from the building blocks, the brain cells, but also how they actually function.

Neurons are the basic units of the brain, just as a person is the basic unit of an organization or a society. Like a person, a neuron is generic and yet at the same time an individual entity. A person changes gradually over time, and a neuron will also adapt. A neuron gradually makes connections across a small gap (the synapse) using an intermediary, a chemical messenger (a neurotransmitter); actual direct physical contact between brain cells is possible but features less. Similarly, a person gradually builds relations with others by indirect contact via a language; touching is rarer. With both chemical messengers and languages there's enormous diversity but also an adherence to the same common principle: communication between two independent entities without any direct physical connection. Both languages and neurotransmitters come in a wide range of varieties, but they can be categorized into families, defined by geographical provenance (for language) or chemical structure (for a neurotransmitter). The actual mode of communication in both cases has parallels in that all languages and neurotransmitters can use a range of signals, from simple to complex and sophisticated. In the most basic scenario, a neuron can signal via its neurotransmitter a simple "yes" or "no," which translates into a momentary inhibition or excitation of the activity of the target brain cell.

When a brain cell "speaks" (or more technically is "active"), it generates a small electrical blip[4] lasting a thousandth of a second (a millisecond), which zooms down to the end of the cell to communicate with the next neuron.[5] But there's a problem once the electrical message reaches the synapse and can go no further. All is not lost, however: the arrival of the blip acts as a trigger for the tip of the cell to release its chemical messenger, which is able to travel across the synapse as readily as words travel through air. Once it reaches its destination, the next cell, the neurotransmitter enters into a molecular handshake with its special target.[6] This interlocking is so tight and tailor-made that a better analogy might be a key fitting into a lock. The complexing of a neurotransmitter with its custom-made target triggers a brief change in voltage in the target cell, effectively a reconversion from a chemical signal to an electrical one. The "yes" in neuronal communication is when there is a momentary increase in electrical activity (excitation); the "no" is when activity is suppressed (inhibition).

Just as most of the time verbal communication is more than a simple monosyllable, with syllables ordered into words, words into sentences, and sentences into a statement, so it is with neurotransmitters: the final effect depends on the sequencing of different neurotransmitters converging over a particular period of time onto a given cell. In both cases the impact of each word or neurotransmitter signal will depend on the wider context over the period within which it occurs.[7] Then, as milliseconds turn to seconds, to minutes, to hours, and eventually to days, the connections effected by this process—the connections between people or between neurons—change.

It's quite fun, and indeed insightful, to explore the various parallels between personal relations and the paths these signals trace through the brain and personal relations: both strengthen through repeated use, becoming stronger and more intense. For both people and neurons, relationships are most flexible when young. Like people, neurons become increasingly specialized and more "individual" as their network grows. Over time, just as people mature and develop particular personality traits, neurons become more resistant to change in general function. And in the same way that friendships wither if they are not actively maintained, underused neuronal connections atrophy.

As an individual grows, he or she establishes more and more complex relationships, some close and frequent, others less activated and more distant; larger and larger groups eventually interconnect and form a still wider society. So it is with the brain, where a nested hierarchy of ever more complex layers of networks of neurons eventually make up a particular macro brain structure. All brain regions eventually interconnect with each other, even over long brain distances, via fiber tracts that operate something like telephone lines, enabling incessant dialogues all over the brain. It is a holistic organization.

The "bottom-up" approach to studying the brain explores just how this organization comes about. If you're a neuroscientist specializing in understanding neurotransmitters, receptors, and how synapses operate, it's a bit like being an expert in interpersonal communication. For example, the neurotransmitter dopamine is linked to many different brain processes, including arousal, addiction, reward, and initiation of movement. But for a bottom-up understanding of how chemicals such as dopamine function, we also need a top-down approach, one that starts with the macro brain areas and attempts to map out how they work together to give rise to different behaviors and ways of thinking.[8] This time an appropriate analogy might be sociology or anthropology, either of which focuses on collective trends and outcomes rather than on the behavior of individuals.

Scientists are now using brain scans to image the wholesale activity of different brain areas as a result of different types of inputs, environments, and behaviors. In a brain scan you might see bright blobs pinpointing certain areas in a sea of gray brain, or perhaps multicolored arrays where white is a hot spot, shading through yellow, orange, and red to a low-activity purple-colored perimeter. But in the enigmatic cohesion of the brain, all the ongoing chatter between the various brain regions will actually *not* be visible to you.

The images of a brain scan reveal the brain at work over a protracted period. Such scans usually have a resolution of seconds (in the very latest developments, tens of milliseconds), but the universal electrical signature of brain cells at work, the action potential, is a hundred or so times faster than that. Brain scans are like old Victorian photographs that show static buildings but exclude any people or animals, which would have been moving too fast for the exposure time. The buildings are perfectly real, but they don't constitute the whole picture.

When looking at brain scans, it is also tempting to think that if a certain area of the brain lights up, it must be the center for whatever behavior or response is being studied. This notion of "centers" of the brain for this or that is attractive: moreover, if it were true, the brain would be so much easier to understand. Back on the cusp of the nineteenth century, Franz Gall introduced the "science" of phrenology (literally "study of the mind"). The white china heads covered with black-lined rectangles labeled with, for example, "love of country" or "love of children" were intended to provide the template against which the bumps of the individual head being studied could be compared to ascertain the strength of a trait. While these busts remain popular with photographers as a prop to enliven shots of media-worthy brain scientists, the approach inevitably was discredited as systematic examination of the brain itself became possible. But traces of the crazy rationale of phrenology, of there being multiple mini-brains within your head, can still fuel interpretations of real scientific findings.

The idea of "one brain area, one function" gained traction as medicine blossomed and clinicians became increasingly skilled at keeping patients alive despite dramatic brain damage from, say, a bullet, an injury, or a stroke. This is where a phrenology-like interpretation was able still to sneak in, by ascribing to the damaged brain area the "function" that had been lost. Yet, as one psychologist remarked more than half a century ago, if you remove a vacuum tube from a radio (yes, the analogy is that old) and the device started to howl, you wouldn't claim the function of the tube was to inhibit howling. If the brain area in question malfunctions, like the elderly vacuum tube, the holistic system of the brain will be impaired, but the contribution of the brain region cannot be extrapolated backward from the final net outcome. To use another analogy, if a spark plug malfunctions, your car will not start, but you can't deduce how a car works by studying a spark plug. We now know that there is no one function controlled by any one brain area. Vision, for example, involves dividing up different aspects of seeing form, motion, and color between as many as thirty different brain areas. And no one brain area has only one function. Rather, each brain structure contributes to a net final function not as a hierarchy but more in the way the various instruments in an orchestra produce a symphony.[9]

This processing in the brain will determine how you see the world, but whatever external inputs are being fed into your brain at any given time, the experience of that very moment *will simultaneously change that organization of brain cells, and hence your thinking.* One leading expert in brain development, Bryan Kolb, sums up: "Anything that changes your brain, changes who you will be. Your brain is produced not just by your genes; it's sculpted by a lifetime of experiences. Experience alters brain activity, which changes gene expression. Any behavioral changes you see reflect alterations in the brain. The opposite is also true: behavior can change the brain."[10] And that is just what we're going to explore next.

HOW THE BRAIN CHANGES

L ondon taxi drivers are renowned throughout the world for their detailed knowledge of the streets, traffic configurations, and one-way systems of the big city. Unlike most of their counterparts around the world, it's seemingly second nature for them to navigate the streets of the British capital without recourse to a map. On average it takes a rookie driver two years to absorb the information required to be able to do this, and to eventually pass an ominous oral exam tellingly called "The Knowledge." These drivers have chosen a career that places a huge burden on their memory, specifically on their working memory, where rules and facts have to be kept constantly in mind in determining ongoing actions.

In 2000 Eleanor Maguire and her colleagues at University College London were intrigued by the question of whether London cab drivers would show any physical changes in their brains as a result of the very unusual daily experience of constantly using their working memory. Amazingly, they saw in brain scans that a particular area of the brain related to working memory (the hippocampus) was actually bigger in the taxi drivers than in others of the same age.[1] Nor was it the case that having a big hippocampus predisposed these individuals to drive cabs, as the difference in hippocampal size was larger the longer the subjects had been plying their trade. This study captured the attention and fascination of the media, as well as of London taxi drivers, of course, and it remains to this day one of the best and simplest examples of the "use it or lose it" principle. Neurons, like the muscles of the body, grow stronger and larger with whatever activity is rehearsed. Even though such adaptation is shared not only by mammals but also by far simpler organisms such as the octopus[2] and even the humble sea slug,[3] humans have been able to exploit this talent superlatively, well beyond any other species.

Changes in the brain as a result of experience were actually first shown as long ago as 1783 by the Swiss naturalist Charles Bonnet and the Piedmontese anatomist Michele Vincenzo Malacarne: they discovered that training dogs and birds led to an increase in the number of folds in a part of the brain (the cerebellum), compared to dog littermates or birds from the same clutch of eggs.[4] However, this finding did little to overthrow the dogma of the time, that the brain was unchangeable, until the idea was revisited in 1872 by the philosopher Alexander Bain: "For every act of memory, every exercise of bodily aptitude, every habit, recollection, train of ideas, there is a specific grouping or coordination of sensations and movements, by virtue of specific growths in the cell junctions." Almost twenty years later, in 1890, the pioneering psychologist William James had a flash of insight: "When two elementary brain-processes have been active together or in immediate succession, one of them, on recurring, tends to propagate its excitement into the other." The actual term for this process, *plasticity*, was first introduced a few years later, in 1894, by the great Spanish anatomist Santiago Ramón y Cajal, who borrowed the word from the Greek root meaning "to be molded,"[5] well before the advent of the ubiquitous synthetic material.

"Give me a child until he is seven, and I will give you the man," guaranteed the Jesuits. Just as plasticity had been anticipated by Michele Malacarne and Charles Bonnet long before modern scientists such as Eleanor Maguire produced experimental data, so too has it been widely accepted that a young, developing brain is more impressionable and more vulnerable. Of course this sensitivity of the young brain to external influence highlights the importance of shaping the right kind of early environment for the next generation. As Hillary Clinton pointed out in 1997, the experiences of children between birth and age three "can determine whether children will grow up to be peaceful or violent citizens, focused or undisciplined workers, attentive or detached parents themselves."[6]

In the first years of life the brain has windows of opportunity, characterized by the exuberant growth of connections between neurons, which allows for astonishing possibilities. For example, in infants the visual and auditory compartments of the outer layer of the brain (cortex) appear to be functionally interchangeable, equally effectively stimulated by either hearing or vision. Consequently, when there is a loss of vision in early childhood, some form of hearing ends up sharper through a process known as cortical remapping.[7] Because the visual sector is not being used for its normal job, it adapts to whatever inputs are available and takes on an alternative role, helping the brain process hearing with a resulting greater prowess.

This obliging adaptation by the central nervous system is not restricted to the senses. One example of the power of the young brain in compensating for damage was the case of Luke Johnson. Luke made the headlines in a British newspaper in 2001 when he was just a toddler. Soon after he was born, his right arm and leg appeared limp and motionless. Doctors diagnosed severe brain damage due to a stroke in the left side of his brain while in the womb or shortly after birth. But within a few years Luke had recovered the full use of his legs and arms. Over the course of the first two years of his life, his brain had been busy rewiring itself, reorganizing nerve pathways to bypass the damaged tissue.[8]

Sadly, these critical periods do not always ensure a positive outcome. Take the case of children who develop cataracts on one or both of their eyes. Visual deprivation through a cataract or another abnormality that impairs sight that occurs between birth and five years, leads to permanent damage to vision. But for children who encounter this problem when they are older, vision typically recovers after treatment.[9] Interestingly, different types of vision have different critical periods, meaning that a child who develops a cataract within a certain time frame may have impairments in, say, the detection of motion, yet develop normal acuity. As with Luke Johnson, the brain of a young child with a cataract will rewire itself, but this time with the tragic consequences that the territory normally used by the nonoperational eye would have been usurped for other purposes.

The notion that there are critical periods of brain development is intuitively easy to grasp, and the changes seen at these particular crucial stages of even normal development are indeed marked. However, it is clear from the remarkable recovery often seen in adult stroke patients that even though "land grabs" in the brain may be less striking later in life, they do not cease with age. In adults as well, various sensory systems can cross the official boundaries between one and another, as when the visual cortex of blind people is activated during the reading of Braille. By the same token, the neuroscientist Helen Neville has demonstrated how auditory impairment induces specific compensation in enhancing vision, while conversely the blind process fast auditory stimulation better.[10]

The same fundamental brain mechanisms driving plasticity during learning in the intact immature brain are also pressed into service during relearning in the damaged or diseased brain. Recovery of function after brain damage falls into three stages: (1) *restoration:* restoring function to the residual brain area, (2) *recruitment:* recruiting new brain areas to aid in the performance of the original function, and (3) *retraining:* training these other brain areas to perform the new function efficiently.[11] With language, the right hemisphere, which is not normally dominant for speech, can take over from the traditional left when it is damaged.[12] Meanwhile, in the case of a nonfunctioning hand in monkeys, just one hour per day of training will keep its neuronal representation in the brain from shriveling to uselessness. This effect has also been demonstrated in humans. Many patients with a malfunctioning hand as the result of brain damage will prefer to use the healthy counterpart, but such a strategy impairs recovery of function. So a sleeve is often placed over the good hand to encourage use of the impaired hand, thereby making it as operational as possible.[13]

The brain does not tolerate "vacant space"—a situation where neurons would not be put to work. The overquoted old idea that we use only 10 percent of our brains is a complete myth, and easy to refute. First, there is no area of the brain that can be damaged without loss of ability of some sort, but if the 10 percent myth held true, we could afford for 90 percent of our brains to be damaged. Second, the brain is the greediest organ of our bodies at rest, guzzling up 20 percent of our energy supplies even though it constitutes only 2 percent of body weight. Why would we use so many resources to maintain 90 percent of neurons to do nothing? Third, brain-imaging techniques reveal that, with the exception of cases of severe damage (such as that seen with a persistent vegetative state), no brain areas show up in scans as completely inactive and silent. Fourth, all brain areas appear to contribute to functions: there is no structure in the brain that doesn't have a job, even though we may not understand exactly how the contributions from different brain areas all fit together to give rise to an ultimate net behavior. Finally, as we've just seen, the brain operates on an unambiguous "use it or lose it" principle when it comes to neuronal survival and connectivity. Were 90 percent of the brain to remain unused, autopsies would reveal large-scale degeneration of up to 90 percent: but this isn't the case.[14]

The harder specific neurons work away at a particular activity, the more brain territory they will take up. In one experiment, Michael Merzenich showed that owl monkeys trained to rotate a disk with two digits only had an enlarged area of the touch (somatosensory) cortex relating to those two digits.[15] This finding has a fascinating counterpart in humans: musicians who play string instruments exercise their left hands more than their right and, in string players, the section of cortex related to touch is accordingly larger for the left hand than the right.[16] Many other examples of plasticity in the sensory system of adults abound, and the impact of repeated experiences on brain functioning are the bedrock of Mind Change, so it's worth getting an idea of just how sweeping and dramatic plasticity can be.

First there are snapshot studies, rather like the one with the taxi drivers, where the brains of a group of people who do something unusual or very frequently on a daily basis show differences compared to the rest of us. Quite generally, for example, brain structures differ between musicians and nonmusicians. Anatomical scans of professional musicians (keyboard players), amateur musicians, and nonmusicians showed size differences in a range of structures: motor, auditory, and visuo-spatial brain regions.[17] It's worth noting that there are strong relationships between musician status and practice intensity, suggesting the anatomical differences are linked to learning and not to a predisposition to music. Meanwhile, substantial time spent doing math induces an increase in gray matter density in specific (parietal) areas of the cortex known to be involved in either arithmetic processing or visuo-spatial imagery/mental creation/manipulation of 3-D objects.[18]

Then there's sport. Experience-dependent plasticity is detectable in the brains of basketball players: when players were compared to healthy controls, there was an enlargement in the brain's "autopilot," the cerebellum.[19] Comparable changes can also be seen in the skilled golfer's brain, albeit in a different cerebral structure, in contrast to those who were less proficient.[20] However, since there was also no linear relationship between a golfer's handicap level and the anatomical changes, it is impossible to say whether the skilled golfers were already predisposed to this particular talent. This chicken-and-egg conundrum is one of the big disadvantages, more generally, of snapshot studies of different groups of people.

An alternative type of experiment that can differentiate cause and effect is to observe changes in the brain over time as normal human subjects with no particular skill or talent are trained from scratch in some standardized experimental task.[21] In one case, it was juggling. Subjects underwent daily training for three months to learn a three-ball juggling task, where perception and anticipation were key to determining upcoming movements accurately. Scans were performed before training, after three months of training, and then after another three months in which no juggling was attempted, by which time performance had deteriorated back to baseline: use it or you will lose it. Meanwhile, the brain scans over this time showed that structural changes occurred within seven days of beginning training and were most rapid during the early stages, when performance level was low. This result suggests that it is the *learning* of a new task that is pivotal in changing the structure of the brain, rather than ongoing rehearsal of something already learned.

Most comforting of all is the observation that such training can still induce brain structure changes in the elderly. In a juggling task like the one just discussed, the performance of the elderly wasn't quite as good as that of a younger population, but gray matter changes *did* occur in identical brain regions.[22] More generally, memory training can induce growth in the cortex in the elderly. When an intensive eight-week training program is deployed, memory performance improves and cortical thickness increases in the experimental group undergoing the memory training.[23] And if older people show brain changes as a result of increased mental activity, it should come as no surprise that younger people do too.

Preparation for the German basic medical exam, the Physikum, can have a demonstrable effect on the brain.[24] This exam "includes both oral and written tests in biology, chemistry, biochemistry, physics, social sciences, psychology, human anatomy and physiology demanding a high level of encoding, retrieval and content recall."[25] Structural changes related to learning occurred in a variety of brain regions related to memory: hippocampus, parahippocampal gray matter, and posterior parietal cortex. But it's not just the acute and stressful experience of exam preparation that's key. Learning a second language increases the density of gray matter, the changes observed being correlated with skill level.[26] Five months of second-language learning, in this case with native English-speaking exchange students learning German in Switzerland, resulted in structural changes that matched up with the increase in second-language proficiency. Once again, the individual amount of learning achieved was reflected in brain structure changes.

The exciting and scary fact of life is that you don't have to actively engage in a specific training task to change your brain: it will happen in any case as a result of the experiences you have and of the environment you are in. In her revealing and fascinating book *The Plastic Mind,* Sharon Begley writes about how "new synapses, connections between one neuron and another, are the physical manifestation of memories. In this sense, the brain undergoes continuous physical change. . . . The brain remakes itself throughout life, in response to outside stimuli to its environment and to experience."[27]

The earliest demonstration of the impact of the outside world was with what was eventually to be called an "enriched" environment and dates back to the 1940s, when the visionary psychologist Donald Hebb did what would be impossible nowadays: he took some of his lab rats home.[28] The actual reason for this bizarre game plan is lost in the mists of time. However, after some weeks in the house, these "free range" rats turned out to have superior problem-solving abilities, such as maze running, compared to the less fortunate counterparts that had remained in standard lab cages.

Since then, more formal studies have shown just how powerful a factor the environment can be, especially when it is stimulating and novel and invites exploration. The very first mention of the term "environmental enrichment" in a scientific article was by Mark Rosenzweig and his team at the University of California in 1964, when they demonstrated for the first time physical changes in neural circuits through experience. The scientists had actually set out to identify the neural mechanisms underlying individual differences in behavior and problem solving in different strains of rats, but they quickly realized the enormous influence that experience had on the behavioral performance relative to their standard caged counterparts.[29]

Over the ensuing decades, neuroscientists have learned that an enriched environment leads to a whole host of physical changes in the brain, all of them for the good: increased neuron cell body size, increased overall brain weight, increased thickness of cortex, greater number of dendritic spines (protuberances on branches of cells that increase surface area), increase in the size of synaptic junctions and hence of connections, and increased number of glial cells (the housekeeping cells of the brain, which ensure a benign microenvironment for neurons). These effects are more pronounced in younger animals but can still be observed in adult or even old rats. There is also increased production of new brain cells in parts of the brain associated with memory and learning (hippocampus, dentate gyrus, and cerebellar Purkinje cells), as well as a greater blood supply and an increase in the amount of growth factors and protein synthesis.

This type of stimulating environment, where there is no fixed task to perform but which nonetheless generates different types of experience, can have a surprising impact even when destiny seems otherwise to be determined strongly by genes. In an experiment done fifteen years ago that has now become a much-cited classic, mice were deliberately genetically engineered to develop Huntington's disease, a neurological disorder that manifests in wild, involuntary movements known as chorea (after the Greek for "dance").[30] The mice left in typical lab cages lived out their genetic fate as they aged, scoring worse and worse each day on a variety of movement tests, while a genetically identical group were exposed to an enriched environment, a world consisting of greater space to explore and more objects (wheels, ladders and so forth) with which to interact. The study conclusively demonstrated that mice living in such a stimulating environment developed movement problems much later and with a far more modest degree of impairment. Even here, with a disorder linked to a single gene and in the less complex brains of mice, nature and nurture interact.

Research since the early 1990s on animals living in an enriched environment have revealed a wide range of physical changes in the brain at the level of individual neuronal networks, as well as demonstrating that the *duration* of the enrichment experience is a significant factor. For example, in one study a single week of environmental enrichment had no effect, but four weeks of enrichment had behavioral effects that lasted two months, while eight weeks of enrichment led to behavioral effects lasting six months.[31]

Given all these physical changes in the structure and chemistry of the brain, it comes as no surprise that animals in enriched environments are superior in tests of spatial memory and show general increases in cognitive functioning such as learning ability, spatial and problem-solving skills, and processing speed. They also have reduced levels of anxiety. In addition, enrichment attenuates the persistent effects engendered by past negative experiences such as prenatal stress or neonatal separation from the mother. The protective effects of enrichment are particularly apparent in animals that are highly anxious or when the task is extremely challenging for the subject.

Enriched environments can also be beneficial in animal models of recovery from brain injury. For instance, transfer to an enriched environment improves the outcome after an experimentally induced stroke, as well as significantly improving motor performance in spontaneously hypertensive rats previously housed in standard laboratory cages, compared with controls remaining in the less stimulating environment.[32] Moreover, an enriched environment will reduce programmed cell death (apoptotic cell death) in the rat hippocampus by 45 percent. And if that were not enough, these environmental conditions can also protect against experimentally induced seizures.[33]

The beneficial and widespread effects of environmental enrichment also persist in aged rats and across a diverse range of species: mice, gerbils, squirrels, cats, monkeys, birds, fish, even fruit flies and spiders—every animal "from flies to philosophers."[34] There is still some controversy as to whether enrichment actually represents a super-special experience or is only a relative improvement over standard laboratory animal housing. However, the main point is that it is the difference between the two types of experience, the relatively greater stimulation, that counts.

But to go back to the question asked at the beginning of the previous chapter: how can an external experience literally leave an internalized mark on the brain? Just as muscle grows with exercise, so too do neurons respond to physical changes, by growing more branches. When it has more branches, a brain cell will have an increased surface area, which makes it an easier target and leads to the possibility of more connectivity with other brain cells. Back in 1949 Donald Hebb came up with the startling suggestion that repeatedly stimulating the same chain of neurons so that they are active at the same time will make them stronger and more effective: as he put it, "cells that fire together wire together."[35] But how exactly? Fast-forward another few decades to when sophisticated techniques became available to monitor the activity of single brain cells (done by inserting microelectrodes inside them and recording the voltage they generate). Using this technology, Swedish physiologist Terje Lomo and British neuroscientist Tim Bliss gained their place in the history of brain research for their breakthrough description of the actual step-by-step process of Hebb's idea. Neuroscientists can now describe the specific physico-chemical steps by which signaling between two brain cells will become more effective as a result of repetition—that is, experience.[36]

While it would be hard to impose a standardized enriched environment on humans, and even harder to justify an experimental "control" group of people deprived of stimulation, the effect of different types of environment has been examined in older healthy adults by investigating the relationship between lifestyle and "cognitive reserve,"[37] namely "the degree to which the brain can create and use networks or cognitive paradigms that are more efficient or flexible, and thus less susceptible to disruption."[38] The findings, perhaps not surprisingly, indicate that a greater involvement in intellectual and social activities is linked with less cognitive decline. It seems that a mentally active lifestyle may defend against cognitive deterioration by increasing the density of synapses (thereby improving the efficacy of communication within intact neurons) and the efficiency of normal and alternative brain networks.[39] Then again, just as in animals, unless the enrichment or stimulation is maintained, performance may decline after previously successful rehabilitation, leading to negative changes. This could be as a result of withdrawal from social situations or reduced levels of activity and/or communication.[40] Even when IQ, age, and general health are all taken into account, older individuals living in a community perform better in cognitive tests than those who are institutionalized.[41]

Most fascinating of all is that even brisk walking may stimulate the production of new neurons (neurogenesis). First, exercise increases the blood supply to the brain, and along with it the all-important oxygen the blood carries. Increased oxygen then enables stem cells (the universal progenitor cells from which different cells

derive) to convert to neurons at maximum capacity, as well as stimulating the release of chemicals that help cells grow. But that's not
all. While physical activity increases the manufacture of neural
stem cells, additional stimulation from an enriched environment
increases the connectivity and the stability of those connections.[42]
Although it has only recently become possible to study cell production in the human brain,[43] changes in brain processes and composition as a result of enriching social, mental, and physical activities
are now thought to help stave off cognitive decline as we age,[44] and
in turn prevent the underlying loss of cells that characterizes the
cycle of death in Alzheimer's disease.[45]

It is also possible for mere thinking to actually change the physical brain, bizarre though this might sound. One of the most-cited
examples of how a thought can drive a physical brain change was
conducted by Alvaro Pascual-Leone and his research group back in
1995 with three groups of adult human volunteers, none of whom
could play the piano.[46] Over a five-day period, the control group
was exposed to the experimental environment but not to the all-
important factor of learning the exercises. A second group learned
five-finger piano exercises, and over the five days showed an astonishing change in their brain scans. But a third group were more remarkable still. The subjects in this group were required merely to
imagine that they were playing the piano, yet their brain scans
showed changes almost identical to those seen in the group undergoing physical practice!

Many additional and amazing examples have followed of the
tangible impact of thinking on the brain. Fred "Rusty" Gage, professor at the Laboratory of Genetics at the Salk Institute, has demonstrated that in order for exercise to generate the production of
new brain cells, the exercise has to be voluntary: the animal must
decide to enter the exercise wheel and run in it.[47] Similarly in humans, it appears that plasticity occurs only when movements are
volitional and/or the subject is paying conscious attention. But if
paying attention at the critical moment is essential for adaptive

changes in the brain, then of still more importance is the individual's state of mind. Perhaps the most familiar but still seemingly improbable example would be the placebo effect, whereby the simple belief that an inert substance has therapeutic properties is sufficient in itself to cure an illness.

We know that this effect works via naturally occurring morphine-like chemicals in the brain, the enkephalins, as research has demonstrated that the drug naloxone, which blocks enkephalins, will correspondingly block the placebo effect.[48] It also turns out that the effects are not merely due to the presence of the enkephalin molecule; rather, it is necessary to believe that the placebo is in fact an active drug. Again, what's all-important is a conscious thought, not just the appropriate bottom-up landscape of brain cells and chemicals.

A further illustration of the key role played by conscious thought can be seen in clinical depression. It turns out that there's a big difference for depressed patients between bottom-up intervention in their condition, with antidepressants such as Prozac, and intervention via various talking techniques such as cognitive behavioral therapy. Psychotherapy differs from antidepressant medication in that the therapist targets the patient's beliefs, encouraging the patient to see the world in a new, more positive way. The cause of the depression—for example, the loss of a loved one—is not diminished but rather is placed in a context that enables the patient to have a more positive outlook. Thus cognitive behavioral therapy for depression works similarly to a placebo. In both cases, the brain is operating from the top down: a belief, which occurs on a macro scale of neuronal networking, which will then trigger chemical changes in the brain, although understanding precisely how this happens is still a great puzzle in neuroscience.

Meanwhile, medication with drugs works differently, by directly modifying from the bottom up. It directly modifies the availability of neurotransmitters, bypassing any personalized neuronal circuitry. And that personalized circuitry, what we can equate with the personal mind, could be all-important. A big difference between cognitive behavioral therapy and direct drug intervention is that the probability of relapse in depression is greater with drugs. Presumably the plasticity changes in personalized neuronal networking shaped by routine cognitive behavioral therapy are more enduring and powerful than a general but essentially transient change in the chemical brain landscape, where drugs are directly manipulating the individual's feelings and conscious state over a much shorter time.

Interestingly enough, in depressed individuals the brain region where new neurons are created from stem cells (the dentate gyrus) shrinks.[49] If these new cells normally would have made it easier to form new connections, Sharon Begley has suggested, then this physical change in the brain might account for why depressed patients are not so receptive to new things, why they persist in seeing the world in an unchanging, unexciting, monochromatic way.[50]

In summary, the brains of a whole range of animals are astonishingly plastic, and the human brain exceptionally so. It is constantly adapting physically to repeated types of behaviors on a "use it or lose it" basis. Such endless neuronal updating is particularly marked in critical time frames during development but continues throughout life into older age. Yet plasticity doesn't stop at the rehearsal of certain skills. The mere experience of living and interacting in a certain environment leaves its mark on the brain, which in turn leads to a unique, personalized brain circuitry (state of mind) that can ultimately lead to further physical changes in the brain and body. But that leaves us with some exasperating riddles. How *can* an insubstantial thought modify a physical state? And, conversely, how can a drug that affects chemicals that modify physical states modify insubstantial thoughts? In short, what is the neuroscientist's story about the possible physical basis of the mind and consciousness?

HOW THE BRAIN
BECOMES A MIND

When she asked me how the brain works, the girl in the audience posed one of the most difficult questions of all. Even before we start to make sense of what all the powerful new neuroscience techniques are actually showing us, we immediately run into a problem with the question itself. After all, what does the phrase "how the brain works" actually mean? The central nervous system carries so many different functions, and on so many different levels of operations, that all this neuronal chicanery cannot really be subsumed under such a catchall single word as "works." For example, on one level, everyone knows how Prozac "works": a key action of the drug is to enhance the availability of a chemical messenger, the neurotransmitter serotonin. But how the increased availability of serotonin "works" to alleviate the subjective misery of depression remains a complete riddle.

Serotonin is, after all, just a molecule; it doesn't have happiness trapped inside it. Instead, the all-important issue is the context, the brain cell circuitry within which it is a bit player—a powerful one indeed, but only when it is operating in the right scenario. Like an actor reciting disconnected lines on her own in an empty dressing room, neurotransmitters and other bioactive signaling molecules accomplish nothing by themselves. They need the other actors, the surrounding scenery, and a clear sequence of events for their lines to have any effect or relevance. In the case of serotonin and depression, we know that there is a lag of at least ten days between beginning to take Prozac and when the therapeutic effects start to be felt. If cheerfulness were a direct product of the serotonin molecule itself, then surely you'd experience an effect immediately upon taking the drug. Having to wait means that the alleviation of depression is not just down to the neurotransmitter itself, its immediate spatial surroundings, or even its direct action on the adjacent cells. Instead, something still more complex is going on within the wider neuronal network, and over a longer time frame.

We've seen that the interlocking of a neurotransmitter with its target molecule is a little like a handshake. Now imagine that handshake persisting, that someone keeps squeezing your hand. Eventually your hand becomes less sensitive, even numb, and more pressure will be needed to achieve the same effect. So it is with the molecular targets. When an individual takes Prozac, the receptors in his or her brain are now going to be bombarded by unusually excessive amounts of serotonin released remorselessly day after day. Slowly the receptors will become less sensitive (the technical term is actually "desensitized"). This suggests that desensitization is a factor in alleviating depression. But how this or any other physico-chemical brain mechanism actually translates into a subjective sensation of either happiness or sadness is one of the biggest mysteries, if not *the* biggest mystery, in neuroscience.

Take another example. Henry Marsh is a distinguished neurosurgeon in London. Many of his operations are conducted while the patient is awake, so that Henry can see the precise functional effects of stimulating the brain in different cerebral locations before any surgical intervention actually takes place. Gory though this might sound, there are no pain sensors in the brain, so it has been quite a routine procedure since the middle of the twentieth century to operate on brains that are fully conscious.[1] However, Henry now has closed-circuit TV in the operating theater and offers the patient the opportunity to watch the whole procedure. Think about it: the brain watching itself. What on earth can be going on?

What is going on, both in Henry's operating theater and in anyone taking Prozac, is an enactment of the "hard problem." This phrase, made famous by the Australian philosopher David Chalmers, refers to our current bafflement as to how the water of brain functioning is converted into the wine of subjective experience.[2] Yet in order to understand how the brain generates consciousness, we need at least some idea, however hypothetical, of what *kind* of answer would work as a satisfactory explanation: would it be a mathematical formula, a brain image, or something more in the realm of science fiction? None of these possibilities seems anywhere near adequate or appropriate. Yet until we know what *kind* of answer we need to solve the hard problem, surely there can be little likelihood of our doing so.

Still, undeterred, some have looked to silicon-based artificial intelligence for an answer. With the ever-growing power of computational processing, the issue here is not so much the intelligence part "I" as it is the artificial part, "A": how would a computer measure up compared with the real biological brain? Many still profess that the brain works "like a computer." This starting premise can be developed in two possible directions: either we can start with biological systems and move toward artificial systems, or we can begin with the artificial and move toward the biological. If we start with a biological phenomenon, be it learning, memory, or even consciousness itself, the usual idea is that we should be able to model it in a silicon-based device. But there's an immediate problem, since the idea of a model requires that we focus on the all-important salient features and jettison the extraneous ones. A model for flight, as exemplified by a plane, requires the defying of gravity; what we don't need is feathers and a beak. So, in order to model consciousness, we would already have to know what the salient physical brain and body processes are, and what bits are extraneous and can therefore be ignored. Yet if we knew that, we would have solved the problem already; there would be no need to bother with the model.

Going in the reverse direction—starting with an artificial system in order to elucidate the biology of cognitive processes, such as learning, memory, or consciousness—can also be treacherous. A distinguished and diverse lineup of scientists such as Ray Kurzweil, Giulio Tononi, and Christof Koch place a premium on "complexity"[3]—that is, in the end it is sheer size that counts in neuronal networks (or, as the philosopher John Searle once quipped, even in a computer made up of old beer cans powered by windmills). In any case, the idea is that if we build machines of ever greater complexity, consciousness will emerge as a spontaneous and inevitable result—and that most-overused of sci-fi characters, the conscious robot, will become a reality.

But this way of thinking overlooks the underlying neuroscience that is normally at work. Consider the trafficking of the huge variety of capricious and subtle compounds in the nervous system that work in different combinations, in different places, over different windows of time, with highly context-dependent and variable effects. The diverse neurochemistry of the central nervous system shows that quality cannot be reduced to quantity, that the complex dynamism of modulating chemicals and our brains is so much more than mere computation.

As we've just seen, neurons are highly dynamic entities capable of extraordinary plasticity, not a fixed component that can be plugged in and played with persistent and dogged regularity, independent of the surrounding micro-, meso-, and eventual macro-scale environment in which it is located. The intense, ever-changing dynamic interaction between coalitions of neurons is nothing like the rigid circuitry of computational devices. No simple systematic accretion of silicon components could ever have the same effect, unless that unit were an exact simulacrum of the neuron, replete with all the chemicals and biochemical dynamics that make possible its characteristic restless plasticity and sensitivity.[4] Moreover, there's a whole body out there, beyond the brain, that receives and sends incessant feedback. Some time ago the neurologist Antonio Damasio pointed out the importance of these chemical signals that feed back and forth between the brain and the rest of the body, chemicals he referred to as "somatic markers."[5] The interplay between the nervous system, the endocrine system, and the immune system—the body's three great control systems—should not be ignored. After all, if they were not interactive we would have biological anarchy; and, even if we didn't, it would be hard to account for the placebo effect, where, as we saw, a thought (namely, some kind of neuronal event in the brain) can impact on health, an event in the immune system.

But just imagine that one day we do develop some kind of artificial device complex enough to be a strong candidate for having consciousness. Let's even imagine that it has passed the Turing test, the hypothetical test devised by Alan Turing, arguably the father of information technology.[6] In this test an impartial observer would not be able to distinguish between the responses of a human and those of the machine. I would still struggle to see how such an artificial system, feat of engineering though it might be, would help solve the hard problem. How might this ingenious conscious computer help us understand how the subjective "feel" of consciousness is actually generated in an objective, physical system? Our inability to determine whether it is a computer or a human answering our questions tells us nothing about the elusive inner state of consciousness: what it is and how it comes about. In any case, it's all hypothetical: the Turing test has still not been passed (although, apparently, there is a human being somewhere who failed it). Whatever their reasons for adopting this approach, for those fixated on building a conscious machine of some sort, perhaps the most exciting goal would be to satisfy the late Stuart Sutherland's criterion: he would accept that a computer was conscious when it ran off with his wife.

Nonetheless, the conceptual impasse of the water-to-wine riddle hasn't stopped neuroscientists, myself included, from trying to make some sort of headway. A zigzag way of progressing is to put the hard problem on hold and instead ask a simpler question: can we lower our sights and just correlate, or match up certain subjective feelings with certain physical events in the brain—say, feelings of well-being with Prozac-induced increases in serotonin—in a way that reveals a consistent relationship between objective events and subjective experiences.

This game plan is the search for what have become known as "neural correlates of consciousness."[7] It's important to note here that no attempt is being made to establish a *causal* link as to how a physical event could give rise to a mental event, or vice versa. A mere *correlation,* just a humble matching up, is more feasible because it sidesteps the conceptual conundrum of the water-into-wine problem. But in order to come up with a convincing correlate of consciousness, we still need a way of describing subjective experience that serves as a kind of shopping list for what we're going to be asking the physical brain to deliver. Yet here's the snag: neuroscience, like all science, strives to be ruthlessly objective and everything we do, all experiments, are painstakingly impartial in their procedures and, most important, they are quantitative, all about measurement.

The catch is that conscious states are quintessentially subjective and qualitative and therefore an anathema to conventional scientists, trained as we are to be impartially objective. So in order to come up with a consistent and persuasive correlate of consciousness, we need to describe subjective states in a way that allows us to draw direct parallels with brain processes. My own suggestion has been to argue that consciousness is not an all-or-nothing phenomenon but is indeed quantitative. Rather than being like a light that is either on or off, I've proposed, consciousness is more like a dimmer switch: consciousness grows as brains grow and develop, both in evolutionary terms across animal species and in individual human development from the fetal stage onward. In adulthood, this variability continues, such that there are times when you are more aware than at others; in everyday jargon we talk about "raising" our consciousness or "deepening" it. In my view, the actual direction doesn't really matter, but we should rather be talking about degrees of consciousness, so we can look in the brain for a physical something, a real process, that also varies in degree from one moment to the next.[8]

As I see it, the most likely neurobiological candidates for consciousness are *neuronal assemblies,* large-scale coalitions of tens of millions of brain cells that can work in synchrony and disband in less than a second. We also know that these highly transient, macroscale phenomena can be dramatically reduced by consciousness-robbing drugs such as anesthetics. The theory therefore runs that the more extensive the assembly profile at any one moment, the deeper the consciousness. In turn, the extent of the assembly at any one time will be dependent on a variety of factors that determine how easily the transient coalition of brain cells can be recruited. One factor would be the sheer intensity of incoming stimulation, which is why an alarm clock will pull you out of unconsciousness into the harsh light of wakefulness.

But then what about when the alarm doesn't ring and you continue dreaming? Here's a situation where there's a weird consciousness of a sort, yet you remain impervious to the external sensory outside world around you. I suggest that the neuronal assemblies that produce dreams are very fragile and not very extensive, since they are driven by the happenstance of internal neuronal activity independent of the strong input of the senses and the external world. And if the assembly in dreams is small, the corresponding consciousness will not be very deep, hence the lack of cause-and-effect logic and the disjointed, improbable narrative that constitutes and characterizes the dreaming state.

If consciousness grows as brains grow, we would also expect this small assembly mode to characterize the mindset of those with still-developing brains: young children, whose behavior is driven by the fleeting moment and instant emotions rather than by step-by-step consequences and planning. Yet there are ways that even the adult human brain could revert to this more basic small assembly mode, despite being fully awake. Many factors in the brain could contribute to the net result that ensues from a small assembly, not just lack of external stimulus (dreaming) or having insufficient brain connections in place (young children). What if there were an excess of a brain chemical that constrained the full spread of an assembly, or what if there were so many sensory inputs bombarding the brain that none had time to trigger an assembly to full potential before it was outcompeted by the next?

I've suggested previously that these two scenarios could occur in schizophrenia and fast-paced sports, respectively, and that in many respects the resultant "small" assembly, which occurs as a consequence of different factors, could nonetheless have a common net state characterized by high emotional content and a momentary consciousness unrelated to past or future.[9] If so, and if the human brain is indeed capable of different modes characterized by different brain states that correlate with different types of consciousness, there will be important implications for the kind of consciousness that might result from continued cyberexperiences. So what we need to do now is explore what normally happens in the human brain as the individual moves from infancy into childhood and then matures into a full, unprecedented human being with a past and a future.

As the great psychologist William James described so beautifully around the turn of the twentieth century, you are born into a "blooming, buzzing confusion."[10] You will evaluate the world around you in purely sensory terms, because all you have are your senses bombarding your brain: how sweet, how cold, how bright, how loud. The wonderful thing about being born a human as opposed to, say, a goldfish, is that although we are born with pretty much a full complement of neurons, it is the growth and connections between the brain cells that account for the astonishing growth of the brain in infancy and early childhood. We've just seen how the generic human brain is capable of very sensitive plasticity that will personalize it into a unique entity, and how a brain cell stimulated by the environment grows more branches, which in turn increase its surface area and thus make it easier to form connections. So we shouldn't be too surprised now that all these available connections can provide us with an adaptability that has important implications for each individual. If you have individual experiences, then you're going to become unique as your particular experiences start to rearrange and reorganize your synapses.

For example, as the weeks go by, connections between a baby's brain cells will slowly grow to accommodate persistent visual patterns of colors and shapes, perhaps consistently accompanied by a particular voice, texture, and smell. And as these connections form, the baby gradually makes the transition from an entirely sensory take on the world to a more cognitive one. Formerly abstract visual patterns and sounds will now be transformed into the baby's mother. And if the mother features again and again in the child's life, then, as with the examples of plasticity we looked at earlier, so will your brain adapt with a unique configuration of brain cell connections and the mother will come to mean something to the child that she means to no one else. Slowly the relation of the child's brain to the outside world progresses from a one-way street to a two-way dialogue. Instead of being constantly in a blooming, buzzing confusion, the child perceives incoming stimuli (a person, an object, or an event) as carrying a meaning wholly specific to that particular child. The child's brain evaluates these stimuli in terms of its existing neuronal connectivity, while at the same time the very experience of doing so further updates the status of those neuronal connections.

In humans, starting at about the age of six years, supernumerary connections—those that are rarely used—start to get selectively pruned back. This is not an impairment, but rather the development of particular patterns of responses and skills that enable the child to navigate and thrive in his or her particular environment. From where the child previously stood at a crossroads, with all possibilities open and unrealized, he or she now begins to take a clear direction, becoming ever more different from everyone else as the brain continues to adapt to each new experience.

Take the example of a wedding ring. It may perhaps first be of interest to a small baby simply because of its conspicuous sensory properties: the gold gleam, the central hole, the smooth round surface that rolls. But as connections associated with the ring become established, the object will slowly gain a meaning as a particular type of object which you put on a finger, eventually further defined as something you put only on one particular finger and only under particular circumstances, then further refined still, as the neuronal connections proliferate, into a broad multifaceted meaning relating to love, weddings, commitment and so on that other merely generic rings do not possess. Eventually, if you acquire a wedding ring of your own, that specific object will have a specific meaning, a relevance that all other otherwise very similar looking rings do not possess. The extensive, highly personalized experiences and hence unique neuronal connections of your brain will have given that object a deep, special significance, "sentimental value," even though in purely sensory terms it is unexceptional. The difference between a generic wedding ring and what might be the most important object in your life is entirely in your head. In this way, the erstwhile one-way street now has traffic going in both directions.

Everything the child experiences from one moment to the next is read against the preexisting associations, but at the same time that current ongoing experience will be updating the connectivity to change it forever. As the child grows, the development of his or her mind will be characterized by this increasingly vigorous two-way dialogue between the brain and the outside world.[11]

So as a child matures, the raw sensation of the outside world gives way to a cognitive take where objects, people, and events have a personalized meaning. But that's not all. Being able to see beyond (rather literally) face value enables a person to evaluate and assess more accurately whatever is happening to him or her. Take the simple case of someone coming into a room on Halloween dressed up as a ghost. While an adult would be able to draw on prior experience and knowledge to interpret the situation as benign, a small child could well be very frightened. Younger children lack the checks and balances of a robust conceptual framework, based on prior experience, that enables them to interpret new events appropriately. Without any frame of reference, however, this strange apparition could be life-threatening.

I suggest that the more we can relate a phenomenon, action, or fact, to other phenomena, facts, or actions, the deeper the understanding. Here is an example. When my brother Graham was only three years old and I was sixteen, I thought it great fun to give him a hard time, as is the way of adolescent elder sisters. One way was to get him to learn by heart great chunks of Shakespeare, and in particular the famous Macbeth soliloquy, "Tomorrow and tomorrow and tomorrow . . ." Graham obligingly learned it like a little parrot and was soon quickly reciting the famous lines on demand, much to the amusement of my giggling school friends. Had I asked him what the line "Out, out brief candle, life is but a walking shadow" actually meant, the best he could have replied would have been something about blowing out the candles on his birthday cake. What he could never have grasped at that age, with his relatively paltry neuronal connectivity, was that the extinction of the candle was really about something else altogether. He could not place the phrase in a wider context and realize that the line was not so much about the extinction of a flame as about the extinction of life—that it was a metaphor for death.

Understanding, then, is basically seeing one thing in terms of another. Surely this is what intelligence is really all about, going back to its literal Latin provenance of "understanding." It is a very different type of ability from the fast processing toward a specified end that, say, IQ tests demand, and which is far more translatable into silicon systems.[12] The mathematician Roger Penrose pointed out long ago that it would be impossible to devise an algorithm for those key human abilities of intuition or common sense. Even further back in time the great physicist Niels Bohr admonished a colleague with the withering put-down "You're not thinking, you're just being logical."

This distinction between the efficient processing of an input to come up with the right output (rote learning of Macbeth, say) and real understanding fits well with a distinction that has been acknowledged for quite a while, that of "fluid" versus "crystallized" intelligence. Psychologist Raymond Cattell first thought up these two distinct concepts back in 1963. Cattell defined fluid intelligence as "the ability to perceive relationships independent of previous specific practice or instruction concerning those relationships."[13] This skill is considered independent of learning, experience, and education. Meanwhile, crystallized intelligence involves knowledge that comes from prior learning and past experiences. Fluid intelligence peaks in our teenage years and then declines, but as we age and accumulate new knowledge and understanding, crystallized intelligence becomes stronger.

This well-established distinction in psychology could correspond directly to whether or not extensive neuronal connectivity is being used. With fluid processing, the efficient input-output processing is free of context, as it was for my brother; there is no need for personalized neuronal connectivity to give a frame of reference. But the crystallized process that is dependent on prior information is an excellent metaphor for extensive neuronal networking. We could even think of the neuronal network structure more literally as a little like a crystalline structure, with intense interconnectivity between the cells. So a neuroscientific definition of the mind would be the personalization of the human brain through its dynamic neuronal connectivity, driven in turn by an individual's unique experiences.

Now let's go one step further. I've often wondered how the unique subjective state of you being *you* is generated at the level of the physical brain.[14] Through the lens of neuroscience, identity is best seen as an activity rather than a state: it's not a solid object or property locked away in your head, but a certain type of subjective brain state, a feeling that can change from one moment to the next. As I see it, there are five basic criteria that the physical brain must deliver in order for you to "feel" that you are a unique entity.

First, you need to be fully conscious, that is, not asleep or anesthetized. And while, as we've seen, neuroscientists still have no objective way of accounting for the subjectivity of each person's unique firsthand experiences of the world, we should not allow this conceptual gridlock to prevent us from moving on to work out further requisites. For example, a rat can be conscious but not have a self-conscious sense of identity. So more is still needed.

Second, your mind has to be fully operational. In the default mode of the normal adult human brain, we've now seen that the mind will enable the individual to react in a certain way to objects, people, and events in accordance with the checks and balances of previous beliefs and experiences. This unique mind, reflected in your unique neuronal connectivity, will enable you not just to make sense of what is happening around you at any given moment, but to make possible the third item on the shopping list.

The third criterion is that you react in a particular way that is determined not just by your past experiences and the prevailing context, as in the second criterion, but also by how those earlier experiences have subsequently shaped your wider beliefs. The key distinction between memories and beliefs is that the former can be evoked independently—a memory gains access into your consciousness without further justification—while beliefs can be appreciated only in terms of how open or resistant they are to validation by potential additional evidence. (I have previously suggested that beliefs could be described along a spectrum, ranging from rational to irrational, in respective relation to this eventual independent validation, or resistance to evidence to the contrary.)[15]

An *irrational* belief (say, all men are superior to women) through to a more *rational* belief (the sun will rise tomorrow) could be so defined in accordance with where they would sit along a crucial single scale of how much they resisted and/or depended on additional evidence. In the brain, this setup could be realized in the extent of neuronal connections and, just as important, by their strength in persisting in the presence of contradictory inputs (say, the obvious prowess of women) that would, could, or should either enforce them or cancel them out, as in the case of sexist beliefs. This potential validation, or refutation, could also be realized, in neuroscience terms, as associations or connectivity that have the potential to offset or cancel out the original association (the belief), but do not actually do so either because the original connection is too strong or the validation is still too weak. These real or hypothetical reactions, your beliefs, in turn, will modify your memories, and how you will respond differently next time around in whatever situations life flings at you.

But there is still more to identity than just having a mind, memories, and even a set of beliefs. Imagine being alone on a desert island. What happens to your identity? On a desert island, who would you actually *be*? I'm suggesting that the issue here would be suddenly not having any context within which to express yourself. The difference between mind and identity is that mind is passive and does not depend on interacting with others, while identity is active and depends on some kind of societal context. Mind is how you see the world, whereas identity is how the world sees you. And for the latter you need a society, a context in which others perceive and respond to what you do. The fourth requirement, therefore, is a context-dependent action-reaction.

Identity in the family, for example, would be inevitably based on strong associations from infancy initially with the colors, sounds, smells, and visual pattern that gradually transform from a conglomeration of raw, abstract senses into, say, the cognitive perception of your mother. Hence in these early years, identity will be strongly linked to momentary consciousness, and will not have much risk of being displaced by any competition, any alternative roles. But as the child grows and other relationships and contexts independent of the family start to develop, identity within the family recedes to become just one of many options and therefore not continuously present in your consciousness. However when that family identity is triggered—say, in the context of Christmas, a wedding, or a funeral—it will come to the fore once again as a strongly dominant identity. Identities that develop later, by contrast, have been laid down over a much briefer period of time, generally much more intermittently, and probably well outside the critical periods of development. Unlike the case of the family identity, the context of the moment will be a much more salient factor in these later identities.

Fifth, this specific instance of action and reaction at a particular moment within a specific context replete with values and memories will now be incorporated into a still-wider framework: a narrative of your cohesive past, present, and future. Thus your subjective awareness of your whole unique life story, captured at any one particular moment but dependent on a hinterland of highly extensive and complex neuronal connectivity, could constitute the moment-to-moment feel of your identity. The scenario of a lifetime of memories and beliefs being funneled into a single moment of consciousness is reminiscent of William Blake's famous lines from the "Auguries of Innocence":

To see a World in a Grain of Sand
And a Heaven in a Wild Flower,
Hold Infinity in the palm of your hand
And Eternity in an hour.[16]

Everything that happens has its own moment in time but can now be linked to all other events by virtue of either preceding or following them. Your identity is therefore a spatio-temporal phenomenon combining the hardwired, long-term, generalized neuronal network of the mind with momentary consciousness, the fleeting generation of macro-scale coalitions of neurons (assemblies) in less than a second. The long-term generalized network of connectivity is your mind, which can now in turn play its part at any particular moment in time. If consciousness is indeed linked to the fleeting generation of macro-scale coalitions of neurons in less than a second, and if the enduring networks of neuronal connections (the mind) can drive a more extensive coalition (assembly), then the ensuing "deeper" consciousness would be directly related to a deeper understanding of events, people, and objects as you encountered them.

The crucial takeaway from this neuroscientific attempt at a deconstruction of identity lies in the vital role of the context in which the mind is operating from one conscious moment to the next. So what happens in situations where this mind is "blown" or "lost"?

8

OUT OF YOUR MIND

I magine a mature, carefully crafted individual brain with connections that are responding to, activated by, strengthened by, and shaped by sequences of specific experiences that no one else has ever had, nor ever will have again. This is the physical basis of an individual's mind. But now imagine those highly individualized connections being slowly dismantled as the branches of the brain cells shrivel. The person would return to a more childlike state, since he or she would no longer have the requisite framework of the adult mind against which to evaluate ongoing experiences. People and objects would no longer have the highly personalized significance so carefully accumulated over a lifetime. We would see the sad and tragic symptoms of Alzheimer's disease, where the patient is indeed "losing" his or her mind, literally *dementia*. Yet we can also "lose" our minds—or, rather, let them go—on a more frequent, temporary, and positive basis, in situations where the lure of here-and-now sensation turns us into passive recipients rather than proactive thinkers.

First, though, a word of caution. We need to be careful not to confuse "blowing" or "losing" the mind, as we will be speaking of it here, with what has been called "mindless" crowd behavior, such as was seen in the Nazis' Nuremberg rallies in the twentieth century, where a collective mob identity derived from political and racial ideologies,[1] just as a collective identity derives from a religious fundamentalism in the twenty-first.[2] In all cases, the overheated and often violent mob is not just blindly emotional, as in road rage or the French *crime passionnel* (where you "see red" and are not accountable for your actions). Far from being "out of their minds," the mob will have a very specific narrative, albeit an utterly repugnant one: they know whom they are targeting in order to enact their revered story line. They are not mindless at all.

If the mind is the personalization of the brain through its individual neuronal connectivity, driven by personal experience, then truly losing your mind would occur when those carefully personalized connections are not fully accessible. For example, drugs and alcohol will impair the chemical communication between neuronal connections, while recreational environments filled with rave music or the rapid-fire stimuli of fast-paced sports do not require a complex cognitive infrastructure, as they are primarily "sensational." Often, the more the raw senses dominate, the greater the pleasure, it seems. The very word "ecstasy" in Greek means "to stand outside" of oneself. It has often intrigued me that we seek out this emotional, unreflective state through diverse pursuits that have one thing in common: an absence of self-consciousness, an abnegation of a sense of self in favor of becoming the passive recipient of incoming senses, indeed of being "abandoned." So you can lose, or be out of, your mind while still being conscious—hence the importance of distinguishing "mind" and "consciousness."

What could be going on in the brain when someone remains conscious yet "blows" their mind? The most obvious tools at the brain's disposal here are the chemical messengers, the neurotransmitters and other modulating chemicals released when neurons are active. One naturally occurring substance in particular is a likely candidate for helping to mediate a sensation-driven experience: the neurotransmitter dopamine. Dopamine is the final common conduit for all psychoactive drugs of addiction, regardless of their primary site and mode of action. The dopamine system has also been linked to processes in the brain relating to feelings of pleasure. For over half a century now, brain scientists have been fascinated by the phenomenon of self-stimulation. Classic experiments by the psychologist James Olds revealed that if electrodes were implanted in certain parts of the brain but not others, rats would work at pressing a bar to stimulate these key brain areas to the exclusion of all else, even feeding.[3] The brain areas that, when stimulated, presumably caused the rats to feel good were those releasing dopamine. In a shorthand but rather inaccurate way, therefore, dopamine has sometimes been simplistically referred to in the popular press as the "molecule of pleasure."

When you are highly excited, are aroused, or feel rewarded—or, indeed, if you are taking psychoactive drugs—this single neurotransmitter will play a key part in delivering all these different subjective experiences. In all these cases, dopamine plays a pivotal role by being released like a fountain from the primitive region at the top of the spine (brainstem) outward and upward throughout the brain, where it then changes the responsiveness of neurons in many different areas. But there is one area in particular that is targeted by dopamine and is of special interest to us here, as it is crucial to human cognition: the prefrontal cortex.

The prefrontal cortex, as its name suggests, sits at the front of the brain behind the forehead. While no one committed brain area is exclusively responsible for making us human, the prefrontal cortex shows huge quantitative differences between our species and other animals. It accounts for 33 percent of the adult human brain but only 17 percent in chimps, our nearest relatives. The prefrontal cortex has more inputs to all the other cortical areas than any other part of the cortex and therefore plays a key role in operational brain cohesion. So if this key area is damaged or underactive, there could be a profound effect on holistic human brain operations.

The classic example of this is the case of one Phineas Gage, who in the mid-nineteenth century was working as a foreman on a railway gang in Vermont.[4] His job was to clear any obstacles in the way of the railway track that was being laid across America at the time. One day, as he was pressing down some explosive material with a large rod referred to as a tamping iron, an alarming accident occurred that earned Gage his place in medical history. The explosive went off prematurely and drove the formidable rod through his brain, more specifically through his prefrontal cortex.

After this terrible event and the reason why the story is now so famous there were, amazingly, no obvious or immediate signs of problems with either Phineas's senses or his movement. Only as the weeks turned to months did it emerge that he had more subtle cognitive problems, such as excessively reckless behavior—not a good trait in someone working with explosives. Surprising though it now seems, Phineas seemed sufficiently unimpaired to return to work, but he had become unbearable as a team player. He was proving to be not only reckless but also, in the words of his physician, Dr. Harlow, "exceedingly capricious and childish . . . [and] particularly obstinate; he will not yield to restraint when it conflicts with his desires."[5] The accident that befell Gage was a living example of a parallel association between an underactive prefrontal cortex and childhood.

In biology a well-known mantra is that "ontogeny" reflects "phylogeny"—individual brain development reflects evolution—so the human prefrontal cortex becomes fully mature and functional only in the late teenage years and early twenties.[6] The years immediately preceding this maturation are what we know as adolescence, which is typified by intensively social behavior, a desire for novelty, and attention seeking, as well as tendencies toward risk taking, emotional instability, and impulsivity. Relationships take on a greater significance, and seeking out fun and exciting experiences becomes a high priority. There is also the likelihood of pervasive negative moods and a feeling of boredom, which may drive the teenager to search for stimuli offering more thrills. Research suggests that adolescents show greater sensitivity to the reinforcing properties of pleasurable stimuli. This may be related to the fact that dopamine production hits a lifetime peak during adolescence.[7] In addition, the teenage years see an increase in the production of another powerful hormone, oxytocin, which enhances feelings of well-being; this may be another factor driving typical adolescent behavior.[8]

Imaging studies of the adolescent brain commonly reveal widespread activity unrelated to any specific task.[9] Such generalized activity decreases as adulthood is reached, implying that the maturing prefrontal cortex becomes better able to coordinate activity and communication across the brain, producing a more organized collection of networks resulting in more efficient processing. As the adolescent brain matures into that of the adult, there is a shift into a more integrated network activity pattern, connecting more distant brain areas; the result is long-range synchronous activity across the brain, enabling improved communication between all the different regions, as the prefrontal cortex is fully operative and thus able to coordinate activity in diverse brain regions.

The subsequent onset of more restrained, inhibitory adult behavior could be due to the fact that the more evolutionarily primitive brain regions (in particular the ventral striatum, which releases dopamine) are fully operational much earlier than the evolutionarily newer ones, such as the sophisticated prefrontal cortex. So teenagers will be more inclined toward risk taking and reward seeking because their prefrontal cortex cannot yet adequately inhibit the brain's more primitive areas.[10]

Adolescents are not the only group who are characterized by an underactive prefrontal cortex and fit this living-for-the-moment profile. Schizophrenia, for example, is the result of a chemical imbalance, and in particular a functionally disproportionate level of dopamine. As a result, the schizophrenic individual's world shifts from the cognitive toward raw sensations driven from the outside.[11] Like children, those suffering from schizophrenia are easily baffled by proverbs such as "People who live in glass houses shouldn't throw stones." Both children and schizophrenics take the world literally, so both might attempt to explain the proverb by saying, "If you live in a glass house, and someone throws a stone at it, your house will break." To them, the external world is a vibrant place that can easily implode on and crush the fragile firewall of the vulnerable inner world.

Yet another, completely different group of those with an unusually underactive prefrontal cortex are those who have a high body mass index (BMI)[12] who are heavy relative to their height. Interestingly, we now know from a recent study using a gambling task that obese people can tend to take more risks.[13]

What could possibly be the common factor between these very different outward states, gambling, eating, schizophrenia, and indeed childhood that have in common an underfunctioning prefrontal cortex?

Anyone who eats knows the consequences of eating too much, and anyone who gambles is always aware of the possible outcome. But the thrill of the moment, be it the sensation of the taste of the food or the excitement of the roll of the dice, trumps the consequences of one's actions in that moment. That is, the brain is operating in small assembly mode, very much like the way it operates while dreaming. The press of the senses, the here-and-now environment, is unusually paramount, as it is for the schizophrenic and for the child. So here are three very different states or activities, overeating, gambling, and schizophrenia, all characterized by an emphasis on external stimulation and an underactive prefrontal cortex: the small assembly mode of consciousness, which we saw just now, could be described as a here-and-now state driven by sensation and, among other things, high levels of dopamine.

If so, then another example of this brain state could also include dreaming, already noted as an example of a shallow, childlike consciousness of the small assembly mode. In fact, a review of imaging studies by Thien Thanh Dang-Vu and colleagues in Liège, Belgium, highlights how dreaming leads to inactivation of the prefrontal cortex.[14] When this key area is underperforming, there is a corresponding drop in holistic coordinated brain operations. Nothing "means" anything, it just is what it is, the small assembly mode of consciousness, where what you see is what you get, and you get it immediately.

Normally, when you are fully awake and accessing your personalized neuronal connections—that is, when you are using your mind—you understand the world in your own special way. For example, the American flag with its stars and stripes may have a profound meaning for a U.S. Army veteran, who carries a highly personalized and extensive network of associations that involve a myriad of events and experiences and incorporate certain abstract values. But for a young child raised in Papua New Guinea, it may be merely a piece of colored cloth with a strange pattern. Your neuronal connectivity, therefore, gives you the ability to appreciate symbolism, to see one thing as standing for something else that could never be guessed from the sensory features of the object alone.

Sometimes we make inappropriate or excessive associations that overinterpret an experience or object, discerning a hidden meaning that to most others would seem neither realistic nor accurate, or even a little crazy. Seeing faces in cloud formations or attributing luck to an object may be everyday examples of such idiosyncratic associations. Similarly, the pairing of two otherwise unrelated events may seem to some to be a silly superstition, but to others it may be a deeply significant sign or portent. Not only do your neuronal connections allow you to imbue objects, events, people, and their actions, with your own personalized "meaning," but they also enable you to understand the world as you live in it. The very act of making these associations, of being aware of a meaning beyond face value, can be regarded as *understanding*. In all cases the person, object, or event is read against your particular neuronal network associations, a conceptual framework that is constantly evolving and expanding as you develop. The more extensive the associations, the larger the conceptual framework in which you could embed the new arrival of the moment and the more deeply you can be said to understand it.

This mind can be distinguished from consciousness, as is evident with dementia patients. Moreover, the various diverse states in which you can "let yourself go" can give clues as to what might be happening in the brain when the mind is not fully operational but you are simply the passive recipient of the senses. We've seen that various extreme states of overeating, gambling, and schizophrenia place an emphasis on stimulation comparable to childhood, and that where the prefrontal cortex underfunctions, most recreational pursuits are also associated with the transmitter dopamine mediating feelings of pleasure. These literally sensational experiences might be characterized by the small assembly correlate of consciousness, a correlate that characterizes non-human animals and the dreaming adult brain, where thought plays less of a role. But how does a thought differ from a raw feeling?

Remember from Chapter 1 the comment "Thinking is movement confined to the brain"? We saw that any thought, be it a hope, a memory, a logical argument, a business plan, or a grievance, has a fixed sequence of cause and effect: a beginning, a middle, and an end. You end up in a different place from where you started. So in physical brain terms, perhaps the basis of thoughts are connections between relevant neurons or neuronal groups. Thinking, that superlative talent of the adult human brain, requires enough neuronal circuitry to take a series of steps and to make connections, along with a correspondingly longer time frame. Meanwhile, emotions can be characterized by their focus on feeling something right now and only now. Conscious thought extends beyond the immediacy of the moment and is not readily trumped by any new here-and-now stimulation.

While information processing is just that, the appropriate response to an incoming stimulus, *understanding,* in contrast, requires that the stimulus be embedded in a conceptual framework. We've seen that a conceptual framework of the type required for understanding can be interpreted, in brain terms, as the growth of the connections between brain cells that are formed postnatally and are subsequently driven, shaped, and strengthened by individual experience. Hence every individual human will have a uniquely personalized brain, as well as a mind that is constantly evaluating the current world in terms of existing associations while simultaneously being updated by it.[15] "Knowledge" would be the embedding of a fact or action within a conceptual framework so that it makes sense, that is, can be understood. "Wisdom" requires still more widespread connectivity, whereby the associations made are drawn from an ever wider range of experience and/or individual memories that enable the assignment of more generalized values.

Table 8.1   Two Basic Modes for the Human Brain?

Mindless	Mindful
Sensation	Cognition
Strong feelings dominate	Thinking dominates
Here and now	Past, present, and future
Driven by external environment	Driven by internal perceptions
Little meaning	Personalized meaning
Not self-conscious	Robust sense of self
No temporal or spatial frame of reference	Clear episodes that are sequentially linked
Children, schizophrenics, gamblers, drug takers; those with high BMI; those engaging in fast-paced sports, sex, dancing, or dreaming	Normal adult life
High dopamine	Less dopamine
Prefrontal underfunction	Normal prefrontal activity
World meaningless	World meaningful
Small assembly correlates of consciousness	Larger assembly correlates of consciousness

As we explore how twenty-first-century technologies drive Mind Change, we will encounter a number of recurring themes, including narrative, a personal life story, and the mind as a real physical entity (namely, the unique configuration of neuronal connections in each individual brain). Table 8.1 summarizes, in an extremely simplified way, how we could think of this mind in relation to the subjective conscious state, as well as various features in the physical brain that we can use as a frame of reference when we come to consider how digital technologies could be impacting not just the generic human brain but individual minds, beliefs, and states of consciousness. We have come a long way from the pink plastic model, but the journey is really only now just beginning.

9

THE *SOMETHING* ABOUT
SOCIAL NETWORKING

People forgetting about my existence is what really gets to me. If I went to a party or on a vacation and didn't document it on my Facebook, did it really happen? Does it just chip away at my presence as a human being and force me to wear an invisibility cloak? . . . I have almost 800 friends on Facebook, but only hang out with a handful of people in real life. Isn't that bizarre? Who are these 790 friends of mine? When's the last time we actually hung out? Do I even know them? If I don't, why would I want them to know me? All of this rhetoric is making me want to simultaneously delete my Facebook and check to see if I have any new messages. Regardless of my decision, I think we can all agree that Facebook has messed with my generation's lives in a very real way. It has dictated our day-to-day lives by creating new social rules and etiquette we must abide by. It's basically turned us into paranoid neurotic messes who are afraid of a real human connection. Mark, why do you have such contempt for us?[1]

This is from Ryan O'Connell, writing in *Thought Catalog* back in May 2011. Although his words are spoken with tongue in cheek, this mindset might be vividly reflecting the colossal impact of social networking sites on our current way of life. If so, is it a sinister sign of a dysfunctional society to come, or does socializing online merely provide a more frequent and accessible version of what all of us have always done? Either way, there will be important implications for our lives and culture in the future. Never before have so many had the opportunity to share music, photos, videos, and opinions as they blog away with ease, and often with almost instant feedback.

While social networks have existed as far back as 1997, sites such as MySpace, Bebo, Instagram, Tumblr, Facebook, Twitter, and LinkedIn remain the most used worldwide, with Facebook dominating the Western social networking market. Compared with other social networks, Facebook users are the most engaged: 52 percent visit Facebook daily, with other popular sites such as Twitter (33 percent), MySpace (7 percent) and LinkedIn (6 percent) trailing behind.[2] The average smartphone Facebook user checks his or her profile fourteen times a day.[3] Thus, while there are numerous social networking sites, much of the discussion here will focus specifically on Facebook, given the popularity of Facebook worldwide and the subsequent amount of research into Facebook use. The "Mark" rhetorically challenged by Ryan is, of course, Mark Zuckerberg, founder of Facebook and *Time*'s Person of the Year 2010. It's hardly surprising that as far as he's concerned, the horizons are unequivocally clear and bright:

> There is a huge need and a huge opportunity to get everyone in the world connected, to give everyone a voice and to help transform society for the future. People sharing more even if just with their close friends or families creates a more open culture and leads to a better understanding of the lives and

perspectives of others. As people share more, they have access to more opinions from the people they trust about the products and services they use. This makes it easier to discover the best products and improve the quality and efficiency of their lives.[4]

I doubt if most people's primary reason for going on Facebook, especially teenagers, will be, as Zuckerberg suggests, the earnest goal of improving the efficiency of their existence. More than a billion people in the world are signed up, and of these just over half visit the site daily.[5] For social networking to be as popular as it is with individuals from such a vast range of cultures and backgrounds, it must be meeting a very basic human need and doing it really well.

The most common reason put forward to explain the immense popularity of sites such as Facebook is that they help us connect online with our offline (real-world) friends and make it easier to maintain long-distance friendships.[6] However, alternative and still popular forms of computer-mediated communication, such as emails or Skype, are effective and easy for communication over long distances. So connecting with friends cannot, on its own, account for the appeal of cybersocializing. Additionally, recent research has found that those who use Facebook to collect a large network of virtual friends report *more* life satisfaction, compared to those who use it to maintain close and enduring real friendships.[7] Alarmingly, this study found that Facebook users are more satisfied with their life when their Facebook friends are regarded as their own personal audience to whom they transmit unilaterally, rather than when they have mutually reciprocal exchanges or more offline relationships within their online networks.

Perhaps it all boils down to that most simple driver of all: the desire to feel good. In one survey, results suggested that the opportunity to develop and maintain social connectedness in the online environment is linked with less depression and anxiety as well as with greater satisfaction with life.[8] Zuckerberg would presumably agree:

> Personal relationships are the fundamental unit of our society. Relationships are how we discover new ideas, understand our world and ultimately derive long-term happiness. . . . We have already helped more than 800 million people map out more than 100 billion connections so far, and our goal is to help this rewiring accelerate.[9]

Already, Zuckerberg is gesturing here at a new type of existence, one in which your identity is no longer so much internalized as externally constructed in close conjunction with others. His use of the word "rewiring" implies that we're functioning together as nodes in some complex machine, that we were already all previously connected ("wired" in a different way), and that this new rewiring is superior. None of these three assumptions is valid. First, although the concept of a global network of thought (the noosphere) was developed, as we saw earlier, by the Jesuit monk Pierre Teilhard de Chardin almost a century ago, it has never been regarded by anyone else as the potential apotheosis of humanity.[10] Second, we have never actually been constantly "wired" together, which is why this novel condition of connectedness is so popular. And third, why should we automatically assume that whatever Facebook offers is superior to all previous forms of communication? We need to look a bit more closely at what's going on.

The antithetical state of being in some way connected to someone else is not to be connected at all, to be alone. In evolutionary terms, there would be survival value and hence a basic subjective pleasure in any behavior that combats solitude. And it turns out that loneliness is really bad for your health. For example, women with fewer social relationships experience strokes at twice the rate of those with more, after adjusting for all other possible factors.[11] Moreover, DNA analysis has identified 209 genes relating to immune system function for combating illness, which are differentially expressed in subjects reporting high levels of social isolation.[12] Evolutionarily ancient immune system defense cells appear to have evolved a sensitivity to socioenvironmental conditions that may allow them to shift basal gene expression profiles in order to counter the changing threats of infection associated with hostile social conditions. Moreover, changes in the expression of inducible genes relate more strongly to the *subjective* experience of loneliness than to objective social network size. And if that weren't enough, loneliness can increase the incidence of cardiovascular disease through reduced levels of oxytocin, the naturally occurring hormone mentioned earlier, which normally reduces and stabilizes heart rate.[13] Because oxytocin surges during close physical contact and is associated with well-being, clearly isolation will inactivate this natural defense mechanism.

The number of people living alone has doubled over the last twenty years; in the United Kingdom, an unprecedented one-third of all adults are in single-member households.[14] This trend is particularly pronounced in the age group twenty-five to forty-four. More people living alone equates with a greater potential for loneliness, so the subsequent arrival of social networking sites will have met a clear demand among a growing group of immediately receptive customers. The subsequent shift in how adults socialize has fundamentally transformed social interaction over the two decades. In 1987, according to one estimate, we spent on average six hours per day in face-to-face social interaction, and four via electronic media.[15] In 2007 the proportion had reversed, with almost eight hours a day spent socializing via electronic media, and only two and a half hours in face-to-face social interaction. The advent of social media not only met an existing need but did so more effectively than normal interpersonal communication. Neuroeconomist Paul Zak has even suggested that social networking itself will increase levels of oxytocin, a hormone normally produced as a result of physical closeness.[16] Perhaps the cybersimulation of being close is the same as the real thing as far as the body is concerned. So what's wrong with that? If we are boosting our oxytocin levels, feeling close to others, and fending off the health-threatening effects of loneliness, what's not to like?

The data on the relationship between feeling lonely and social networking are surprisingly complex.[17] Research shows that people who actively engage in Facebook via messaging friends and posting on friends' walls report lower levels of loneliness than those who primarily engage in passive observation of friends' profiles.[18] People who report feeling lonely also apparently are more strongly emotionally attached to Facebook, which indicates that it is the more solitary who use the site to compensate for their lack of offline relationships: meanwhile those with healthy, already established real-life networks simply turn to Facebook as something additional that is nice to have.[19] Interestingly, students with higher levels of loneliness also report having more Facebook friends than those who are in reality more sociable.[20] Thus, while social networking might be used to deal with feelings of loneliness, it may not have the desired effect after all. For example, the futurologist Richard Watson has serious reservations:

> I believe that one of the main reasons that Facebook and Twitter are so successful is that we are lonely. . . . Universal connectivity means that we tend to be alone even when we are together. You can see this when couples go out to dinner and spend most of their time texting or when kids get together for play-dates and end up sitting next to each other on separate gaming consoles for hours on end.[21]

Some researchers suggest that escaping online to avoid real-world problems may actually exacerbate them.[22] One study examined Facebook use from the perspective of adult attachment theory, which emphasizes the role of the primary caregiver during infancy.[23] Attachment theory was developed by psychiatrist John

Bowlby in the mid-twentieth century, when he was treating emotionally disturbed children. Bowlby proposed that attachment
could be defined as "lasting psychological connectedness between
human beings," and he showed that babies were either "secure,"
"anxious," or "avoidant" in their attachment styles.[24] The secure
baby might cry when the mother left the room but would start to
play again as soon as she returned. In the case of anxious babies,
however, when the mother came back, they would push her away
and burst into tears. In contrast, the avoidant baby would act as if
nothing had happened, despite a rise in heart rate and levels of the
stress hormone cortisol.

Adults behave like babies too. While secure people feel comfortable with intimacy, avoidant individuals struggle to establish emotional connections. Avoidant individuals are more likely to be
socially isolated and to attempt to shut down their emotional needs
in relation to others. In contrast, anxious individuals worry about
being alone; they fear rejection and will engage in behaviors that
they believe will strengthen their relationships. The researchers
found that individuals with high levels of anxious attachment used
Facebook more frequently, were more likely to use it when feeling
negative, and were more concerned about how others perceived
them on Facebook.[25] So it would seem that Facebook fills a need for
those with maladaptive early experiences. However, it's still unclear whether Facebook use could help those with high levels of
anxious attachment by combating feelings of loneliness and reinforcing their relationships.

But it's not just the lonely and the anxious who are drawn to
social networking. Research has also shown that individuals with
higher levels of openness spend more time on Facebook and have
more friends there.[26] Openness signifies an active imagination, a
willingness to try new experiences, an attentiveness to inner feelings, a preference for variety, and having a curious mind. Thus,
having a large number of Facebook friends is, paradoxically, associated both with higher openness levels and also with being more

lonely. Although it might seem counterintuitive, openness and loneliness are not incompatible: openness is a personality trait, whereas loneliness is a state. A combination of the "pull" of wanting to be open and the "push" of loneliness is a potent factor in determining just how much you give away about yourself. It is this self-disclosure that is crucial to understanding the real appeal of social networking sites.

As a species, we seem to have such a craving for self-disclosure that it could be considered a very basic part of the human psyche. Harvard scientists have actually demonstrated that sharing personal information about oneself, as on social networking sites, activates the reward systems in the brain the same way as food and sex do.[27] Astonishingly, the participants in this particular experiment were even willing to give up monetary rewards for the opportunity to talk about themselves. The results also suggest that the existence of a reciprocal cyclical feedback for self-disclosure rewards and perpetuates the sharing of personal information on a basic biochemical level. Consequently, the appeal of social networking is rooted in a biological drive of which we are largely unaware and which we find difficult to control voluntarily.

Although we may not be aware of it as a basic, biological need, the conscious craving for personal expression and self-disclosure could be the key to what so many find compelling about Facebook and other types of cybersocializing. Although social networking sites will, of course, make such communication easier, the socializing itself may not be the key issue. Instead, the *real* hook may be the experience of transmitting personal information on an unprecedented scale, because Facebook and other comparable sites encourage you to divulge information about yourself to others in a way you may never have done before. When someone updates her status with something personal, she shares it with her hundreds of Facebook friends. Just think about it. Of course we have shared personal information with one another since the dawn of time, but now we do it with 262 people (the average number of Facebook friends across all ages and demographics) instead of just a few close friends.[28] The point is that, when you share personal information on Facebook, whether through your profile or as a status, you share it with an immediate audience that is the largest ever in human history.

If so, then the next question is, why are we willing to give away so much personal information on such an unprecedented scale? Perhaps the rewards of participating in social networking sites and the psychological disposition toward self-disclosure reinforce each other. One of the most consistent outcomes of computer-related research shows that the lack of face-to-face communication leads to a corresponding rise in self-disclosure, because we don't have visual cues or access to the appropriate body language to discourage us from self-disclosing or to make us second-guess what we disclose.[29] When we meet people in the flesh, shake their hand, look them in the eye, and pick up on cues through body language, we gradually build trust and rapport; we come to feel we know the other person before we let our guard down. Until then, defensive body language, averted eyes, physical distance, and tone of voice may all act as warnings not to give too much away too soon. Body language is an ancient evolutionary mechanism that signals us when we should let our defenses down and when we should maintain them. If there are no such cautionary signs, nothing to prevent us from talking or writing on and on and on, then disclosure is far easier. People who want to disclose more will use social networking sites more, which in turn only encourages them to disclose even more.

For example, 488 users of social networking sites were surveyed in Germany twice within a six-month period.[30] Individuals with a stronger disposition to self-disclose showed a higher tendency to participate in such sites. At the same time, frequent social networking use increased the wish to self-disclose online, because self-disclosing behaviors are reinforced through accumulating social capital within Facebook and similar environments. The $64,000 question then is: why? If loneliness is the main driver of social networking use, there are far more effective, reciprocal, and personal ways to communicate with individuals than the ubiquitous status update online. Yet the lonely are the most attracted to the screen. Just why is it so pleasurable (as the Harvard study clearly demonstrated)[31] to divulge your feelings and thoughts not to a single confidant occasionally but to an audience of hundreds or thousands on a daily or even hourly basis?

Arguably, with time and distance to hide behind, you can portray yourself as someone completely different and more interesting. The opportunity to avoid the awkwardness of hesitating and stumbling over your words seems wonderful, especially as you won't have a chance to say anything you don't mean or might regret. You feel secure and inviolate as you derive tactile pleasure from tapping the keys, and you see the writing on the screen dance to your precise command and control. Another part of the excitement of being online comes from being constantly connected. Someone somewhere is always available to interact with you right now; after all, you are globally wired. But at the same time, you can say anything you like without the embarrassment or discomfort of a face-to-face interaction. No wonder such an experience makes you feel good.

In 2011 a joint Italian and American investigation aimed to dissect the type of experience people have while using Facebook.[32] Is it primarily a relaxing experience or a stressful one? Thirty students aged nineteen to twenty-five took part in short exercises in which they first looked at panoramic landscapes (the relaxing experience), then spent three minutes navigating their own Facebook account, and finally spent four minutes completing a stressful task, such as solving a mathematical problem. During these tests, their physiological stress levels were recorded to measure how stressful or relaxing the participants found each trial. During the stressful experience, activation of the fight-or-flight system was triggered, resulting in increased respiration, sweating, and pupil dilation, whereas the relaxing experience led to activation of the parasympathetic nervous system, which resulted in the opposite reactions. What was most interesting was that navigating one's own Facebook page appeared to offer an experience that was neither relaxing nor stressful, but a more active positive state. Participants showed a mixture of physiological responses also seen in the relaxing and stressful conditions. The researchers concluded that the success of social networking sites "might be associated with a specific positive affective state experience by users." In short, going on Facebook is physically and/or physiologically exciting. But what biological process actually triggers this experience of feeling good, of enjoying Facebook more than you would, say, looking at a painting or going for a walk?

We saw previously how brain scientists have long been fascinated by the phenomenon of "self-stimulation," where rats will spend all their time working at pressing a bar to stimulate key brain regions, to the exclusion of all else including feeding. The areas that, when stimulated, presumably caused the rats to "feel good" were those releasing the transmitter dopamine. As well as contributing to feelings of pleasure, dopamine plays another role in the diurnal rhythms of sleep and wakefulness, where it is linked to heightened alertness. Just think of the hyperactivity caused by "speed" (amphetamine), which releases abnormally high levels of dopamine in the brain. It's not difficult to see an overlap between feeling excited and feeling happy. Many activities in life that are arousing, such as fast-paced sports, are also rewarding. Suffice it to say that if various brain states relating to arousal and reward are consistently linked to raised levels of dopamine, and if social networking is rewarding and exciting, it is very likely that social networking might serve as another trigger for the release of dopamine in the brain.

Dr. Susan Weinschenk, a behavioral psychologist who has published five books on user experience in computer systems, has listed the specific features of Facebook and other social networking sites that might make them triggers for dopamine release.[33] First, they provide instant gratification: you can now connect to someone immediately and probably get a response in a few seconds. Second, they offer an anticipatory thrill. Neuroimaging studies show higher stimulation and activity when people anticipate a reward than when they actually get one.[34] Similarly, the anticipation of whatever new tweets, updates, or comments on your profile you might find drives your fascination with social networking sites more than the actual information you receive. Third, these sites offer small pieces of information. The dopamine system is most powerfully stimulated when the information coming in is modest enough not to satisfy entirely. The limited capacity of a tweet or a "like" is therefore ideal to activate the dopamine system. Finally, there's *unpredictability.* This is the much-studied reward/punishment mechanism involved in intermittent or variable schedules of reinforcement. When you check your email or text, or use Twitter or Facebook, you don't know exactly who is contacting you or what you'll receive. This feedback mechanism is largely unpredictable and exactly what stimulates the release of dopamine in your brain. The posting and receiving of entries on Facebook or Twitter could trigger the release of small blips of dopamine, possibly encouraging such activity to become compulsive.[35]

This almost instant feedback from others, which is unlike any real-world interactions, is much more prevalent when there are so many more people out there in cyberspace who can oblige. The sight of a name flashing up presents a little burst of excitement, a little blip of dopamine that will ensure anticipation for the next fix; you can never actually be satiated. But then why should the mere sight of a response on your particular site, irrespective of what it actually says, trigger that blip of dopamine in the first place?

Attention and approval from adults are among the strongest rewards we experience as we are growing up. Infants need a meaningful relationship with a caring and involved adult in order to survive, grow, and thrive. Astonishingly, human growth hormone is

thought to be released in proportion to the amount of caring attention a child receives.[36] When babies cry to announce their hunger
or other discomforts, they rely on the world, particularly adults
nearby, to correct the problem. These demands are necessary for
survival, and when they are met the existence of the child is acknowledged. A hungry baby who yells until someone comes with
the right source of nourishment *knows* that he has an effect on the
world. The world acknowledges that he exists. This tiny human already has significance. A baby whose needs are ignored eventually
gives up and "ceases to exist." In extreme—and, fortunately, very
rare—cases of neglect, such infants stop crying when they are hungry and literally starve to death. A child's emotional well-being begins with attention paid to his basic physical needs. Yet the need
goes further: the caregiver has to approve and to show approval.
Once the physical needs are met, this drive for further validation is
one of the strongest motivating forces in our nature. When we
aren't met with positive feedback, we no longer feel safe and protected. And over time we become conditioned to crave approval
not just from our parents but from others as well.

The importance of such recognition does not diminish with
age. Unlike the real world, Twitter and Facebook can always be relied on to provide an almost instant response to even your adult
demands for attention. Facebook may readily be filling a gap that
friends and family do not fill so comprehensively.[37] This in turn
may explain why the obsessive social networker relies on the illusion of cyberintimacy, despite the inevitable price of a loss of privacy. Many of us take privacy for granted until we feel it is being
invaded, whether by an intrusive personal question or the extreme
scenario of a helicopter from Google Maps hovering outside the
bedroom window. As film star George Clooney quipped: "I don't
like to share my personal life . . . it wouldn't be personal if I shared
it."[38] Until now, most of us most of the time have felt in control of
our private lives—of how much we confide, to whom, and when.
But now such assumptions no longer hold.

It is impossible to give an operational definition of privacy, but most of us, until now, have had a strong instinctive sense of it. In his first nonfiction book, *The Blind Giant,* novelist Nick Harkaway weighs up the balance between the blessings and the threats of the Internet:

> Privacy is a protection from the unreasonable use of state and corporate power. But that is, in a sense, a secondary thing. In the first instance, privacy is the statement in words of a simple understanding, which belongs to the instinctive world rather than the formal one, that some things are the province of those who experience them and not naturally open to the scrutiny of others: courtship and love, with their emotional nakedness; the simple moments of family life; the appalling rawness of grief.[39]

In contrast, at a technology conference in 2010, Mark Zuckerberg defended his controversial decision of the previous year to change privacy settings that pushed users to reveal more personal information, saying, "We decided that these would be the social norms now and we just went for it." Zuckerberg told his audience that Internet users don't care as much about privacy anymore: "People have really gotten comfortable not only sharing more information and different kinds, but more openly and with more people and that social norm is just something that has evolved over time."[40]

Already privacy appears to be a less prized commodity among the younger generation of Digital Natives: nearly half of teenagers have given out personal information to someone they don't know, including photos and physical descriptions.[41] Meanwhile, over half of young people send out group messages to more than 510 "friends" at a time (the number of Facebook friends an average youth has),[42] fully aware that each of these contacts could then pass on that information to *their* network of hundreds more. The trade-off for more attention and the possibility of fame is, and always has been, loss of privacy, and it was always a tough call on how to achieve the appropriate balance. So how is it that we previously treasured privacy so much, yet now hold it in increasingly casual disregard? Until now, privacy was inextricably linked to an internally generated sense of identity; the one always entailed the other. But if identity is now constructed externally and is a far more fragile product of the continuous interaction with "friends," it has been uncoupled from the traditional notion of, and need for, privacy.

Of course, for many, social networking is a fun adjunct to a normal life that enhances communication with existing friends made in the real world. Yet there is more to the popularity of these sites than their trendiness and ability to make life easier would suggest. Social networking sites could be viewed as a kind of junk food for the brain: harmless enough in moderation, but deleterious when you overindulge. It seems that the *something* about social networking harnesses and promotes a potentially vicious biochemical cycle, whereby evolutionary biological forces ensure that humans feel good when they are combating loneliness by sharing personal information with others, mediated by the release of dopamine in the brain. As a result, self-disclosure creates a hit of pure pleasure as direct as that derived from food, sex, dancing, or sport. Until now, this natural urge to let it all hang out has been counterbalanced by the rigors and constraints of body language in face-to-face communication, which makes you all too aware of your private self. This awareness of being a private individual can serve the very valuable role of ensuring that we are not manipulated or taken over from the outside. So, by constraining the natural urge to disclose information about ourselves to everyone and anyone, the opposing desire for privacy will ensure that only trusted individuals access the "real," vulnerable you.

However, social networking removes these constraints, allowing individuals to disclose more through this medium than ever before. The consequent trading in of the age-old birthright of privacy may mean that others will think less of the "real" you that is now revealed. But imagine if this mode of constant self-disclosure and feedback became the norm. It might become increasingly difficult to protect the "true self," with all of its weaknesses and failures, from being reshaped and supplanted by an exaggerated, ideal self that is presented to an audience of hundreds of "friends" and "followers." So what would happen if such a cyber-airbrushed persona started to elbow out the real you?

SOCIAL NETWORKING
AND IDENTITY

"Over the next 10 years, people's identities are likely to be significantly affected by several important drivers of change, in particular the rapid pace of developments in technology."[1] So reads the opening salvo of *Future Identities,* a report commissioned by Sir John Beddington, at the time chief scientific adviser to the British government. His starting point was that "the emergence of hyperconnectivity (where people can now be constantly connected online), the spread of social media, and the increase in online personal information are key factors which will interact to influence identities." Is this all just scaremongering by a high-profile establishment figure, or is it a serious and urgent wake-up call?

Social networking sites evolved from the 1990s version of the Internet, which was already providing many new ways to communicate and socialize. Computer-mediated communication back then was dominated by forums, early online games, chat rooms, bulletin boards, and so on, all of which had a default setting of anonymity; it was up to the user to make it personal.[2] Individuals logged on and could select any name they wanted to use as an alias—for example, John_Smith9000. Those who've researched this earlier style of computer-mediated socializing have suggested that it was this potential for anonymity that was all-important: it allowed individuals to discover their repressed identities and learn more about themselves, presumably in a fairly safe way.[3]

Thus, initial investigations into online self-presentation mostly focused not so much on identity but rather on its absence in anonymous or pseudonymous online environments. These investigations found that individuals tended to engage in role-playing games and unusual behaviors in an environment that was arguably healthier than that of the real world.[4] In contrast, nowadays anonymity is no longer an inherent part of socializing online. The interesting question then is what happens when you are "nonymous" (i.e., not anonymous) in an online environment.[5] The identities that result are very different.

Technology researchers Nicole Ellison and danah m. boyd (who prefers her name in lowercase) have defined present-day social networking sites as sites that enable a user to (1) "construct a public or semi-public profile within a bounded system"; (2) "articulate a list of other users with whom they share a connection"; and (3) "view and traverse their list of connections and those made by others within the system."[6] Revealing personal information is now part of setting up a social networking profile: Facebook requires a user's real name.[7] While there are always ways around this, the point is that social networking sites have transformed computer-mediated communication by tethering it to your real-world identity. Additionally, a significant proportion of a user's "friends" are people that they know or have met in real life. That's a massive and important shift: socializing on the Internet has become fiercely personal. Identity is therefore a central issue, as are shifting notions of identity in relation to social networking sites.

But how *you* see yourself need not be shared with everyone else. Your online self and your "true self" are not necessarily the same.

This notion of a "true self" was first introduced as long ago as 1951 by the influential American psychologist Carl Rogers, widely considered to be one of the founding fathers of psychotherapy.[8] His theory was that the true self is based on existing characteristics that need not necessarily be fully expressed in normal social life, perhaps because there are not necessarily occasions when they'll be manifest; rather, they are imagined as particular reactions in hypothetical situations. Fifty years later, the digital age saw John Bargh and his team developing the concept of the "true self on the Internet" to refer to an individual's tendency to express the "real" aspects of the self through anonymous Internet communication rather than face-to-face communication.[9] The idea is that the Internet provides individuals with a unique opportunity for self-expression that encourages people to reveal their true self, including the aspects that are not comfortably expressed face-to-face. Because of this effect, cybercommunication could be regarded as more intimate and personal than face-to-face communication. Those who form friendships in this way through social networking sites are more likely to put a premium on self-disclosure, in the hopes of expressing their true self.

According to Katelyn McKenna at New York University, people who believe that they are better able to express their true self on the Internet are more likely to form apparently close relationships in cyberspace.[10] Moreover, people with a stronger tendency to express their true self in this way in the cyberworld are more likely than others to use the Internet as a social substitute.[11] Using the Internet as a social substitute involves establishing new relationships with strangers and having Internet-only friends. Such people are more likely to develop a compulsive passion for their Internet activities.

In a survey of university students that explored their motives for Facebook use, a particularly interesting result was that individuals with a strong tendency to reveal their true self on the Internet reported using Facebook for establishing new friendships and for initiating or terminating romantic relationships more frequently than individuals who were less concerned about expressing their identity.[12] So it would seem that, for some though not all, the use of Facebook as a vehicle for self-expression goes hand in hand with it being their main conduit of friendship. Wanting to express one's true self through Facebook is also linked with obsessive Facebook use.[13] Once again there is a paradox: those who most keenly desire to express their "true" identity are precisely those who rely most heavily on relationships in cyberspace. So it's not so much that Facebook is inherently good or bad, but rather how it is used, and the role and importance it plays in one's life.

Unlike in the real world, a Facebook identity is implicit rather than explicit: users show rather than tell by stressing their likes and dislikes instead of elaborating on their life narrative, their strategies and attitudes for coping with problems and disappointments, and all the other baggage of a normal life.[14] Someone who posts a picture of a chocolate cake without any meaningful accompanying explanation leaves it up to her audience of "friends" to infer what they will. In a real-life relationship, the cake might be a physical tie-in to a much deeper and personal story: it could bring back fond memories of a shared excursion or the sense of triumph that comes with mastering a new recipe. But without shared associations—special common experiences or interests—the cake will "mean" nothing. The same could apply to people. As one Facebook user, a female student I spoke to, described it:

> When you get to know people on Facebook whom you've hardly met, you may think at first that you know them; but it turns out that you only really know the artificial things, bands and movies they like—you'll not know how they react to situations and crises in a way that reveals their "real" identity to others and even to themselves.

The most interesting question, however, is this: might this new and different way of expressing yourself actually mean that you see yourself differently?

Whether or not a social networking profile expresses a distorted "true" self or displays something more comparable to the real self, there is no doubt that whatever identity a person is most comfortable promoting, it is likely to be the best possible version. Untagging unflattering photos and deleting regrettable posts are just two examples of micromanaging the types of information to be seen by colleagues, family, and friends. Unsurprisingly, looking good in a photo is the most important factor reported by teenagers when considering which profile picture to select on a social networking site.[15] Canadian sociologist Erving Goffman described how in general human beings are always on the alert to how others react to us, continuously adapting our outward demeanor to ensure the best possible image.[16] Goffman died in 1982 and therefore never lived to see the advent of Facebook and Twitter. Yet he understood how we long to promote our "front stage" self, while the real "backstage" self pedals away furiously to ensure the most impressive performance. These are desires that sites such as Facebook and Twitter now cater to superbly by providing the widest audience ever.

Adapting this dichotomy of front stage versus backstage to the Facebook culture, we can think of a "networked identity," a term first coined by danah boyd, who described it as follows:

On MySpace for example, you have to write yourself into being: in other words, you have to craft an impression of yourself that stands on its own. Is it the end-all and be-all in developing your sense of self? Of course not. But online expressions are a meaningful byproduct of identity formation.[17]

Research is showing that the identity portrayed on Facebook is neither the uninhibited real self previously displayed in anonymous computer-mediated environments nor the self presented in three-dimensional, face-to-face interactions.[18] Rather, it is a deliberately constructed, socially desirable self to which individuals aspire but which they have not yet been able to achieve.[19] Astonishingly, social networking has now resulted in three possible selves: the *true self,* expressed in anonymous environments without the constraints of social pressures; the *real self,* the conformed individual who is constrained by social norms in face-to-face interactions; and, for the first time, the *hoped-for, possible self* displayed on social networking sites.[20]

But perhaps this is splitting hairs. It turns out that that there is little difference between how an observer rates the personality of a Facebook page owner based on displayed material and the Facebook page owner's actual traits.[21] Nonetheless, the possibility of online identity management allows for distortion. Researchers agree that, like a funhouse mirror, the online self is likely to be an exaggerated version of the real self. And this exaggeration could get out of hand. It's not that social networking sites have provided the first-ever opportunity for us to distort our identity and hence relationships, but they are now providing an unprecedented opportunity to do so. Creating, managing, and interacting through an online profile is a chance to advertise yourself unchallenged by the constraints of reality, to be an idealized, edited version of the "real" you. Although this online self is "an invention that, for most people, is a continual approximation of presenting our sense of self to the world,"[22] the clinical psychologist Larry Rosen fears that a dangerous gap could grow between this idealized "front stage" you and the real "backstage" you, leading to a feeling of disconnection and isolation.

One direct outcome could be an exaggerated obsession with the self, since many researchers have commented on how social networking sites provide the ideal platform for narcissists.[23] Given the extent of control one has over one's online presentation and the scope of audience that can be reached, a bidirectional relationship might come as no surprise. Social networking can demonstrably increase narcissism levels. In the meta-analysis mentioned earlier, Jean Twenge and her colleagues investigated more than fourteen thousand college students and found that twenty-first-century students scored substantially higher on questionnaires for narcissism compared to those from twenty years earlier.[24] However, Facebook use did not become widespread until after 2006, which means that in this study any screen-based effects on indulging the ego would have to be attributed to earlier forms of social networking sites. True, but Facebook could now be tapping into this existing predisposition (which is another reason for its popularity), thereby feeding the trend for self-obsession in a self-perpetuating cycle.[25]

This relationship between heightened narcissism and social networking, though well documented,[26] appears to be confounded by a number of different factors, such as the number of friends, status updates and photos, and the types of interactions with other users. The connection needs to be unpacked further, as narcissism itself can be unpacked. Narcissism turns out to be a complex phenomenon that can be broken down into a range of characteristics: exhibitionism (showing off), entitlement (believing that one deserves the best), exploitativeness (taking advantage of others), superiority (feeling that one is better than others), authority (feeling like a leader), self-sufficiency (valuing independence), and vanity (focusing on one's appearance).[27]

Research shows that adults who score high in superiority have a preference for posting on Facebook. For the younger generation of students, it is posting on Twitter that is associated with superiority, with Facebook activity linked to exhibitionism.[28] In contrast, for adults, Facebook and Twitter are both used more by those focused on their own appearance, but not as a means of showing off, as is the case with college students. These complex findings are important, as they reveal just how many factors are involved in different types of social networking and in the very different groups of people who are all users. Most interesting of all for Mind Change is the generational difference between students and adults, which suggests that a lifetime of early exposure to the influences of Facebook and Twitter is producing a cultural mindset that is different from that of previous generations.

But what remains true across different age groups, and irrespective of the particular characteristics that predominate in the group, is that enthusiastic use of social networking sites is linked strongly to narcissism. Of course, human beings have always been vain, self-centered, and prone to bragging, but now social networking provides the opportunity to indulge in this behavior unabated and around the clock. Interestingly enough, such behavior may also be linked to low self-esteem.[29]

For those of any age with an existing network of friendships built up in the three-dimensional world, social networking sites can be a happy extension of communication, along with email, Skype, or phone calls, when face-to-face time together just isn't feasible. The danger comes when a fake identity is both tempting and possible through relationships that are *not* based on real, three-dimensional interaction, and/or when the most important things in your life are the secondhand lives of others rather than personal experiences. Living in the context of the screen might suggest false norms of desirable lifestyles awash with friends and parties. As ordinary human beings follow the activities of these golden individuals, self-esteem will inevitably plummet; yet the constant narcissistic obsession with the self and its inadequacies will dominate. We can imagine a vicious circle where the more your identity is compromised as a result of social networking and the more inadequate you feel, the greater the appeal of a medium where you don't need to communicate with people face-to-face.

Individuals with low self-esteem perceive Facebook as a safe, appealing place for self-disclosure, and they spend as much or more time using Facebook than people with high self-esteem.[30] A world of airbrushed online portraits may appear to be a low-risk environment ideal for enriching relationships by sharing things they might otherwise be too inhibited to share. However, people with low self-esteem tend to post updates that emphasize their negative features at the expense of the positive, compared to those with high self-esteem. As a result, they are "liked" less than people who have a higher opinion of themselves.[31] When asked the reasons why people unfriended others on Facebook, 41 percent nominated the annoying status updates as reasons.[32] Ironically, therefore, the conviction that it is safe enough to disclose their feelings on Facebook may encourage people with low self-esteem to reveal things that lead to the very rejection they fear.

Moreover, given that the majority of a Facebook user's "friends" do not spend time in face-to-face interactions, the impression many such "friends" have of someone with low self-esteem is likely to be negative, and this leads to further rejection.[33] In contrast, expressing insecurities in face-to-face interactions typically takes place with a close friend in a trusting and intimate fashion. In contrast, the unique platform of social networking sites can lead other users to perceive as distasteful the negativity of an effective stranger with low self-esteem. This creates a situation where Facebook contact may be the only way that many "friends" communicate, yet people with low self-esteem who "overshare" on Facebook will, ironically, deter others from becoming close to them.

While many see Facebook as a harmless tool for maintaining existing friendships, a recent study found that avid users attach too much significance to the type and amount of attention they receive on their Facebook page, and hence are disappointed.[34] The conclusion is depressing:

> Facebook appears to be a tool for transforming both close connections and unknown others into audiences for individualistic self-displays. . . . [P]ublic self-displays on social

network sites may be one way young people today enact increasing values for fame and attention. . . . [N]ew communication technologies augment an individualistic focus on the self.[35]

Data from both self-report and observer rating show that individuals are more likely to express more positive emotions and present better emotional well-being on Facebook than in real life.[36] Moreover, Facebook can open up an alternative world in which people can escape from reality and be the person they would like to be. We are also being exposed to "perfect" lives as we read about people who seem to have it all and are always smiling. These apparently wonderful lives increase the pressure on us to be perfect, admired, and fulfilled: a goal that is inevitably doomed to failure. Perhaps it's more than a curious coincidence that, over the last twenty years, the number of people saying there is no one with whom they can discuss important matters has nearly tripled.[37] In summary, the culture of social networking may predispose users to a narcissistic mindset that in turn enforces low self-esteem. By relying on Facebook to satisfy this need for approval, users not only think less and less of themselves but also long desperately for others to notice and to interact with them. This in turn encourages the development of an exaggerated or completely different identity: the hoped-for, possible self.

Although this scenario may seem far-fetched, it is precisely what might now be happening. Kidscape, a British charity that helps prevent bullying and protects children, conducted a survey in which they assessed young people's cyberlives through an online questionnaire.[38] Of the twenty-three hundred or so respondents between ages eleven and eighteen taking part from England, Scotland, and Wales, one in two say they lie about their personal details on the Internet. Of those, the one in eight young people who speak to strangers online are the most likely not to tell the truth, with

60 percent lying about their age and 40 percent about their personal relationships. This suggests that many young people adopt a different identity online. Although this particular survey was concerned with children's safety online, it also highlighted the fact that children often create a different persona when they interact with others, especially strangers, in a way that they wouldn't or couldn't in the real world. The survey found that young people start to change their identities and to act differently online at just eleven years of age; create identities that allow them to be more rude, more sexy, more adventurous; and generally indulge in inappropriate behavior. However, knowing that people may be viewing your input and judging you accordingly could encourage young people to edit their material and be overly self-conscious. This new trend might well be just harmless fun, but then again it also might herald a society where relationships are based on ephemeral connections between imaginary identities.

Social networking sites seem to be enabling, for the first time, some kind of unreal, idealized self—in the words of a twenty-one-year-old girl, an "alter ego." People sometimes actually talk of a split personality, an online self as opposed to an offline self, like a Dr. Jekyll morphing from time to time into a cyber Mr. Hyde. For Mr. Hyde there are no constraints on behavior and new possibilities are therefore opened up beyond the mere "fun" Dr. Jekyll could ever have just being himself.

As far as the brain is concerned, it is impossible to disentangle identity from environment and context, as we've seen. So it is inevitable that the identity of the next generation will be formed in the context of a pervasive, ever-changing cyberculture. The very structure of our lives means that friendships in the real world face competition from those we develop as we turn to constantly present and convenient social media. For those who do not have persistent and stable relationships, obsessive indulgence in cyberfriendships might have a negative effect on identity. Most worrying would be the dominance of the "front stage" mentality of living primarily to gain approval and recognition in the eyes of others, where whatever you might be doing is assessed as to whether or not it's Facebook-worthy. There is the risk that those with impressionable minds and relatively little experience of the real world may become overly concerned with their social lives and define success or achievement in terms of how many friends they have on Facebook or followers on Twitter.

There is even the suggestion that social networking maps directly onto the physical brain: Professor Ryota Kanai of University College London has claimed that the size of an individual's online social network is closely linked to certain aspects of physical brain structures implicated in social cognition.[39] Specifically, the team found that variation in the number of friends on Facebook strongly and significantly predicted the size of certain brain structures. They also found that the gray matter density of one particular brain region, the amygdala, was linked to social network size in the real world, and also correlated with the extent of a subject's online social network.

But what does this scientific-sounding result actually tell us? Could it really be the case that using social networking sites can change brain structure, or that those who already have a certain brain structure will have a larger online social network? The difficulty lies not so much in what the scans themselves show but in the danger of overinterpretation. However fascinating this study may be, a simple brain scan does not tell us whether an activated area is an effect, a side effect, or even a cause of the behavior being observed. The imaging of different brain areas is excellent for a correlation between brain and behavior, but it does not mean that that area is the center *for* that behavior. The light on your iron does not mean that it's the center for the function of the iron; it is just a corollary, a side effect of the iron working.

Remember that brain regions don't have single jobs that map one-to-one onto behavior in the outside world. Apart from the most primitive brain regions, such as the specialized cells control-

ling respiration, the more sophisticated areas of the brain participate in many different functions. There is no big boss or hierarchy
of command. So what does it actually mean when a particular brain
area is comparably larger or denser in a scan? Interpretation, and
the validity of that interpretation, will depend very much on just
how precise the activity is that is being matched up to the scan.

Think back to the London taxi drivers exercising their working
memory of the streets of London and how that corresponds with
changes in the size of different brain areas, as shown in scans. The
skills involved in knowing the best routes for navigating the metropolis are far more specific and definable, and less vague, than
those in forming friendships. And again, the rookie pianists who
imagined playing the piano were nonetheless still performing in
their mind a specific set of movements, whether or not the actual
contraction of muscles then actually ensued. A network of friendship is a far more abstract concept and thus harder to define operationally.

Still, we shouldn't throw the neuroscience baby out with the
bathwater of simplistic interpretations. Instead, let's think about
the complex ways in which the delicate, malleable brain responds
to social networking, from the instant a pulse of dopamine is triggered by a response to the latest tweet through the long-term shaping of brain cell connectivity, which will ultimately result in a
lifelong rearranging of synapses in the brains of those who might
eventually be regarded as narcissistic or low in self-esteem.

Sherry Turkle has laid out a convincing case in her book *Alone
Together* for the paradoxical-sounding argument that the more connected you are, the more isolated you feel.[40] If you are constantly
connected, you are a kind of commodity that can be compared to
others and found wanting. This scenario was described in Oliver
James's *Affluenza* in relation to material goods and a dysfunctional
lifestyle in a capitalist society: if you believe that you need to be
more beautiful and richer than the next person in order to have
significance, and if you see other people also as commodities for

enhancing your perceived significance still further, you will be incapable of having the kind of human relationship essential for well-being.[41] Each person is reduced to a series of check marks in boxes, with no independent worth despite being in a constant state of comparison. It is precisely these qualities of connectivity and comparison that have come to define the quintessence of social networking.

Social networking sites provide an unprecedented platform for social comparison and envy.[42] One 2013 study that investigated the relationship between envy, life satisfaction, and Facebook use found that Facebook had triggered more than 20 percent of all reported incidents of envy or jealousy. Primarily caused by self-comparison with the social lives or vacations of others, this envy subsequently decreased life satisfaction. However, since previous research has indicated that most individuals portray an exaggerated or falsified state of contentment, the result may be a vicious cycle of portraying exaggerated happiness, feeling envious of others' happiness, and experiencing a subsequent need to increase the portrayed levels of one's own well-being.

This cyclical arms race driven by the basic brain mechanisms of addiction and reward would be a far cry from the identity and narrative of a life story that has until now given us our purpose and which mandates an elaborate cognitive context developed throughout life. That is not to say that envy and unhappiness, which are part of our cognitive makeup, don't interact with the biological hook of the dopamine cycle. They have to. And if so, paradox though it might seem, are we becoming weirdly addicted to constant comparison with others, even if it ultimately makes us unhappy? Perhaps the unhappiness, that flat, let-down feeling of disappointment, is simply because you didn't win this time around, so try again; spin the wheel or roll the dice again and next time you just may be lucky and impress everyone else. And if you could do that, it would mean you were "cool."

So what defines "cool" on social networking sites? In the past, status was proclaimed by your watch, your car, your achievements. Now status for the Digital Native is measured not by possessions or a prestigious job, but by how "famous" (loosely defined) you can be. Interestingly enough, "coolness" has now been democratized. Wealth, gender, and age are no longer relevant. Achievements are no longer required. It's simple networking that counts. Those who decide to keep only close friends on their Facebook profile may lose out in another way, since the number of friends one has on Facebook is seen as related to one's physical and social attractiveness.[43] (Just in case you need reassuring, the optimal number of friends in regard to social attractiveness has been found to be 302.)[44]

For those seeking a quick and painless way of combating low self-esteem and promoting the self, a San Francisco–based company named Klout could be the answer. It provides social media analytics to measure a user's influence across his or her social network. The analysis takes data derived from sites such as Twitter and Facebook and measures the size of a person's network, the content created, and how other people interact with that content. The result is a Klout score that reflects your online influence.[45]

In case you're thinking that a Klout score would be an irrelevance when it comes to the mainstream real world, ponder the following disturbing comment from a recent article: "Just as an SAT score is used to judge students and a credit score is used to judge financial standing, [Klout creator Joe] Fernandez hopes that the Klout score will become an 'ingredient' in job interviews."[46] Since he's the founder, perhaps his predictions are a little biased and overly enthusiastic. Still, Klout makes me feel queasy. First, according to Klout, impact is based entirely on the activities carried out on social networking sites; second, it is the quantity rather than the quality of your messages that is evaluated; third, the response you generate provides an opportunity for you to use your "influence" to draw attention to different brands. People may receive "Klout Perks"—free products or discounts—based on their online impact.

Although Klout denies you'll have any obligation to talk about the product, the possibility of receiving perks such as free laptops and airline tickets even if you don't have a high Klout score means that your friendships have become advertising space. And the fact that importance is measured through social networking, that it depends on how much attention you attract, and that this attention can be rewarded is unlikely to bring out the best in anyone. What kind of lesson are you learning about relationships—and, indeed, about how you see yourself?

For some with robust experience of real-life relationships, spending time updating social networking sites and communicating with friends may improve well-being, just as a good gossip session on the phone might, but there is the danger that "well-being" could now be achieved simply by being "popular" among other Facebook users or by having a high Klout score. While in the short term well-being is obviously a good thing, if in the longer term you begin to question such a superficial reason for feeling happy as a high Klout score, you may begin to feel that something is still missing from your life, such as the sense of fulfillment typically gained from hard work, a real-life challenge, a sporting achievement, or creative skills. In any event, taking things to the extreme, consider this question: how would any of us feel living in a future society where the end point for achieving a feeling of contentment was simply the sheer number of people noticing you in cyberspace?

"I deleted Facebook," a friend from the real world confided, "because it was just like high school all over again, where every girl is more popular and beautiful than you are." While some individuals may be ready to break this false happiness cycle altogether, they remain in the vast minority. In 2011, a hundred thousand U.K. Facebook users deleted their profiles.[47] In a study of Facebook quitters, the main reason cited was privacy concerns. Individuals with higher Internet use were more likely to quit Facebook, indicating that they had been concerned about their obsessive social networking.[48] The very fact that quitting has been termed "virtual identity suicide" by social networking researchers indicates the importance some place on their Facebook profile.

When we were looking at the neuroscience of identity, I suggested that it entails the carefully constructed and unique mind interacting with a large number of momentary external contexts over time. Those contexts and that interaction will be hugely significant in determining who you are and how you see yourself. Until now, the adult mind was the product of a dialogue between environment and self, and this dialogue allowed for pauses, self-reflection, and the slow but sure development of a robust internal narrative. In contrast, an unremitting environment lived out on social networking sites will present the polar opposite: a scenario that displaces a robust inner sense of identity in favor of one that is externally constructed and driven. And because such an identity would be so strongly dependent on the responses of others, it would recapitulate the insecurity and fragility of a child's lopsided, still-nascent sense of self.

Until now, the continuing dialogue between the individual and the environment has been weighted in favor of an internalized, personalized life story and inner commentary that, I've suggested, amounts to what we call identity. As we've just seen, the very basic drive to share this narrative with other people has traditionally been offset by the biologically based constraints of face-to-face interaction, where friendships are formed gradually and in a highly selective manner. However, social media removes these evolutionary precautions and presses the accelerator on unfettered self-disclosure in a context where the usual brakes applied by normal interpersonal feedback are absent. So instead of a small circle of friends, the self is now publicized to an audience of hundreds— and, like all public performances, it is held up to endless scrutiny and comment. How will this overly self-centered yet fragile identity fare in interpersonal communication and relationships?

SOCIAL NETWORKING
AND RELATIONSHIPS

E ven in ancient Greece, the importance of face-to-face interaction over mere words on a page was recognized. Socrates
warned: "Every word, when once it is written, is bandied
about, alike among those who understand and those who have no
interest in it."[1] Nowadays, the screen provides the opportunity for
abandoning interpersonal interaction on an unprecedented scale,
and with that abandonment comes a wholesale reduction in the
risk of embarrassment and feelings of unease in social interaction.
No one can see you blush, hear your voice go squeaky, or feel your
damp palms. But then again, nor can you pick up on these all-
important cues for working out how the other person might be reacting.

In 2012, the British communications watchdog Ofcom produced its ninth annual Communications Market Report. The director of research for Ofcom, James Thickett, was acutely aware of the
significance of the decline that year's report found in the number
of mobile calls, which dropped by 1 percent, and in the number of
landline calls, which decreased by 10 percent. He concluded:

In just a few short years, new technology has fundamentally changed the way that we communicate. Talking face-to-face or on the phone are no longer the most common ways for us to interact with each other. In their place, newer forms of communications are emerging which don't require us to talk to each other, especially among younger age groups. This trend is set to continue as technology advances and we move further into the digital age.[2]

Ofcom reported that the average person was now sending fifty texts a week.[3] A staggering 96 percent of young people ages sixteen to twenty-four were using instant message (non-oral) communication—email, text message, or a social network—every day to contact friends and family. Meanwhile, verbal communication over the phone or in person has become correspondingly less popular, with only 63 percent talking face-to-face with friends or family daily.[4]

Although Digital Natives may prefer non-oral communication through text messaging or the Internet, the type of emotional support that can be provided by these forms of communication turns out to be very inferior. Researchers at the University of Wisconsin–Madison asked the following question: could the content alone of an emotionally supportive conversation between a parent and a teenager convey reassurance, or would the tone of voice and/or physical presence of the parent also play a role?[5] In the experiment, teenagers performed a stressful task, and were then comforted by their parents over the phone, in person, or through instant messaging, or had no parental contact whatsoever. Salivary levels of cortisol (a marker of stress) and oxytocin (an indication of bonding and well-being) were measured afterward. Teenagers who spoke with their parents over the phone or in person released similar amounts of oxytocin and showed similar low levels of cortisol, indicative of a reduction in stress. In comparison, those who instant-messaged their parents released no oxytocin and had salivary cortisol levels as high as those who did not interact with their parents at all. Thus while the younger generation may favor non-oral modes of communication, when it comes to providing emotional support, messaging appears comparable to not speaking with anyone at all.

The extent to which this increase in communication online is not just a symptom but a cause affecting young people's ability to socialize and empathize in face-to-face conversations has not yet been empirically established. Such reluctance to make human contact with someone, especially a stranger, may be the product of a fear of, or simply a lack of practice in, this most basic of human talents. However, neither alternative bodes well for society. Imagine that you've never had much practice at face-to-face communication because, from an early age, you've interacted with others mostly through a screen. Instead of body language, tone of voice, and physical contact, the dominant vehicle for expression is words. It's hardly surprising that many people complain that they've been misinterpreted when chatting through social media. However much you may discuss your emotions, the statements just cannot live up to true facial expressions.

Scarier still is the idea that real, nonverbal communication might be subverted by a parallel cyberuniverse in which the skills of interpersonal interaction are not sufficiently rehearsed; if they are not rehearsed, it is unlikely that you will be any good at them. So perhaps many younger people, brought up with the safer option of communicating online, prefer not to risk looking them in the eye, giving a hug, or taking a chance that their voice may rise up an octave. In turn, this might mean that online relationships are indeed very different from real ones. Professional matchmaker Alison Green has found she faces unique problems when dealing with Digital Natives: they appear to struggle to communicate face-to-face, and have shifted the development of romantic relationships online, with couples preferring to get to know each other first through the distance and safety of their smartphones.[6]

The big question is whether such a trend is to be welcomed or not. Sherry Turkle has suggested that Facebook gives "the illusion of companionship without the demands of friendship."[7] But in a relatively recent review, Paul Howard-Jones concludes that, all in all, the Internet "can benefit self-esteem and social connectedness."[8] Reaching a similar conclusion, Moira Burke from Carnegie Mellon University surveyed over a thousand English-speaking adult Facebook users for two months around the world, recruited through an ad.[9] The results showed that Facebook increased bonding and decreased loneliness with direct communication. But, tellingly, as users passively consumed news they felt they had less access to new ideas being generated by others. Most important of all, loneliness was experienced in proportion to the amount of content they consumed. These findings highlight a possible crucial difference between actively supporting existing friendships and the passive consumption of other people's social news. Favorable outcomes from relationships on social networking sites *appear to apply only to those communicating with existing friends.* It turns out that using the Internet to make new friends actually has a very different result. A long-term study of the relation between adolescent boys' and girls' computer use and their friends and quality of friendship reveals that using the Internet to make new friends is now linked to lower levels of well-being.[10]

Along similar lines, drawing from a sample of preadolescents and adolescents, researchers found that online communication was positively related to the closeness of friendships.[11] No surprises there. However, this effect held only for respondents who communicated online primarily with already existing friends and not for those who communicated mainly with strangers. It was the socially anxious respondents who perceived the Internet as more valuable for intimate self-disclosure, and this perception in turn led to yet more online communication. So it seems that real-world social intimacy and Facebook intimacy are far from being the same thing—a distinction borne out by a recent survey.[12]

This crucial dissociation between the number of cyberfriends and real-life emotional ones also applies to older generations. This time a study examined the relationships between use of social media (instant messaging and social network sites), network size, and emotional closeness in individuals ranging in age from eighteen to sixty-three.[13] Perhaps not surprisingly, time spent using social media was associated with a larger number of online social network "friends," but it was *not* linked to larger offline networks or with feeling emotionally closer to offline network members. So, in general, how might online socializing differ fundamentally from that in the real world? One difference might be in the development of interpersonal communication skills, and consequently in empathy.

The ability to care about and share others' emotional experiences is something that clearly differentiates humans from most of the rest of the animal kingdom.[14] Studies have found that even babies and toddlers show empathetic behavior. One investigation with thirty-four-hour-old infants showed that even very young babies cry to the sound of another newborn's cry, and that the cry is a response to the vocal properties of the other's cry. The babies exposed to the crying of another newborn cried significantly more often than those exposed to silence or those exposed to a synthetic newborn cry of the same intensity.[15]

However, full-flowering empathy is not necessarily guaranteed as part of our birthright. It would be hard to imagine a complex trait such as empathy being completely a product of our genes. For example, although work by Ariel Knafo and his team at the Hebrew University of Jerusalem indicated a significant genetic contribution—indeed, an array of genes will inevitably be necessary for the realization of the diverse cognitive traits of the healthy human brain—the actual ability to empathize with others keeps maturing well into our twenties.[16] So there is ample time for the environment and the experience of relationships to play a significant part in determining our ability to empathize.

The term "emotional intelligence" has increasingly crept into everyday language to define the "ability, capacity, skill, or a self-perceived ability to identify, assess, and manage the emotions of one's self, of others, and of groups."[17] Whether or not emotional intelligence is part of or different from more general intelligence is an interesting question—but not our immediate priority here. Suffice it to say that if it's something that, like intelligence itself, varies from person to person, then emotional intelligence cannot be a feature that is determined at birth. As mentioned in Chapter 4, a survey of fourteen thousand U.S. college students suggests that levels of empathy may be declining.[18] While this survey, like all surveys, cannot provide a causal link between the soaring popularity of social networking sites and the decline in empathy, the somewhat eerie correlation is undoubtedly worth considering.

A particularly interesting approach by Miller McPherson was to compare ideas of friendship in 1985 with those in 2004. McPherson's team discovered that the 2004 subjects had fewer people they could really talk to, with the number of available confidants down by about a third. Even more alarming, the proportion of those having no one at all with whom they could discuss important matters had nearly tripled.[19] While there were losses from both within the family and in friend groups, the largest deficits in confidants occurred in the community and neighborhood. McPherson and his colleagues raised the possibility that respondents might have interpreted the question as pertaining to strictly face-to-face discussion, and if so, the shift from oral to online communication may account for the apparent decline.

It is easy to see how these two trends—a decrease in empathy and an increase in online relationships—could be linked. As the psychologist Larry Rosen has pointed out, if you hurt someone's feelings but cannot see the other person's reaction, you'll lack sufficient cues to understand what you've done and apologize or take some other compensatory action.[20] The increase in feelings of isolation may be connected with the ease and speed with which personal information can be posted, which may encourage people to thoughtlessly send potentially damaging information out into the world. And if empathy arises from the experience of interpersonal face-to-face communication, but we are good only at what we rehearse, then the reduction in face-to-face communication would reduce our ability to empathize. Empathetic connections in the real world could be a good analogy for the networking between individual neurons that occurs in the brain (recall Hebb's famous words about neurons: "cells that fire together wire together"). However, if you have no one whom you feel cares about you, you might be all the more tempted to be uncaring to others or just care less about being so. And what effect might this indifference have on our own view of what is important and appropriate to share?

Beyond empathy, excessive Internet use could lead more generally to a reduced ability to communicate effectively, as it has been associated with a lack of emotional intelligence, including poor performance in interpreting facial expressions.[21] Perhaps it is unsurprising that people who spend excessive amounts of time on the Internet have deficits in face processing. One particular study used a visual detection system to compare the early stages of the processing of face-related information in young excessive Internet users by analyzing their EEGs.[22] By presenting subjects with images of faces and objects, researchers had discovered that the brain waves elicited by the viewing of faces were generally larger and peaked sooner than similar responses elicited by objects. This meant that the faces had more significance for the average observer than the objects. However, excessive Internet users generally had a smaller brain wave response than normal subjects, whether they were looking at faces or at tables. This result suggests that for heavy Internet users faces were of no more importance than everyday inanimate objects. Although it's still unknown whether these impairments would extend to deeper processes of face perception, such as face memory and face identification, these observations indicate that excessive Internet users have deficits in the early stage of face-perception processing, an impairment that is in turn associated with a range of disorders including psychopathy and autism.

In the United Kingdom alone, more than half a million people—around 1 percent of the population—have a form of autism. Autism spectrum disorders are characterized by a triad of impairments: (1) difficulty with social communication, both verbal and nonverbal, such that patients often find it hard to "read" other people; (2) difficulty recognizing or understanding other people's emotions and feelings, as well as expressing their own; and (3) difficulty with social imagination, namely, understanding and predicting other people's behavior, making sense of abstract ideas, and imagining situations outside their immediate daily routine. Traditionally, autistic spectrum disorder is diagnosed within the first two years of life. Hence some specialists claim that it is impossible to link autism to social networking, since very young children won't be accessing such sites. Yet Dr. Maxson McDowell, a psychoanalyst, has pointed out that individuals who obsessively use social networking could still develop autistic-*like* traits, such as avoiding eye contact. In infants, early eye contact initiates the ability to connect with others' subjective experiences that is so essential for social communication and interaction, an ability that is impaired in autism.[23] Indeed, infants' inability to track their mother's face is often associated with a future diagnosis of autism.

Meanwhile, three academics at Cornell University, Michael Waldman, Sean Nicholson, and Nodir Adilov, have explored possible associations between technology use and the later development of autism. They considered a variety of screen activities, including watching television, watching videos and DVDs, watching films in a movie theater, and using a computer. A link emerged between early TV watching and autism. If TV can be a factor, it would hardly be surprising if the screen world of the Internet turned out to have an impact as well.[24]

So if we accept the broadening of the term "autistic-like trait," the Cornell findings might suggest that we shouldn't exclude environmental factors in some cases. Rates of autism diagnosis have been increasing rapidly in the past two decades, and the increase cannot be attributed solely to genetic causes. One study by Irva Hertz-Picciotto and Lora Delwiche at the University of California, Davis, showed that even after taking into account changes to diagnostic criteria and the broadening of the autism spectrum, a significant proportion of the rise in autism cases was still unexplained.[25] We should not dismiss out of hand the possibility that there are triggers in the environment, such as prolonged and early exposure to the world of the screen, where no one looks you in the eye. Humans have an evolutionary mandate to adapt to their environment, and when that environment does not provide opportunities to rehearse the interpersonal skills essential for empathy, one result might be the development of autistic-like difficulties with empathy.

Interestingly, David Amodio at New York University and Chris Frith at University College London have shown that one of the symptoms of autism is an underactive prefrontal cortex.[26] Recall from Chapter 8 how essential this brain area is in ensuring that the brain functions cohesively. If this key area is underactive, there could be a profound effect on holistic brain operations, creating the mindset described earlier in which the sensory trumps the cognitive and nothing "means" anything: it just is what it is. A laugh, a frown, a blush, a smile might "mean" a lot less: what you see is what you get simply at (almost literally) face value.

Whether or not screen technologies could ever increase the possibility of autistic-like behaviors, it is well accepted that the reverse holds true, and autistic people are generally most comfortable in cyberspace. Catrin Finkenauer and her team at the University of Amsterdam investigated the link between autistic traits and Internet use in a longitudinal study and showed that people with a tendency toward autistic traits, especially women, were more prone to compulsive Internet use.[27] This evidence suggests some kind of link between an attraction to the Internet and impairments in empathy, as was also seen in the study of heavy Internet users' lack of distinction between faces and objects.

On the positive side, autistic individuals' affinity for the screen has been already exploited for therapy. One notable example is the U.K.-based ECHOES Project, which helps schoolchildren with autism experiment with difficult social scenarios. ECHOES is:

> a technology-enhanced learning environment where 5- to-7-year-old children on the Autism Spectrum and their typically developing peers can explore and improve social and communicative skills through interacting and collaborating with virtual characters (agents) and digital objects. ECHOES provides developmentally appropriate goals and methods of intervention that are meaningful to the individual child, and prioritizes communicative skills such as joint attention.[28]

Why should the screen hold such appeal for someone who has problems with empathy? The most obvious answer is that in such a world, there is no need to understand what might be going on inside the minds of others—what you see is what you get. Given the absence online of all the valuable nonverbal cues we have been discussing, perhaps we are all autistic-like when we go online.

To sum up: there is a link between atypical brain wave responses in problematic face recognition, characteristic of autism, and also of heavy Internet users; a link between autistic spectrum disorders and an underfunctioning prefrontal cortex, indicative of a more literal take on the world; a link between early screen experiences and later development of autism; and a link between autistic conditions and the appeal of screen technologies. While it is impossible to establish cause and effect between these various links, and indeed to draw any firm conclusions, there appear to be some parallels between heavy Internet use and autistic-like behaviors that deserve further exploration. This line of thinking inevitably brings us to question what we mean in any case when we talk of a relationship. Surely, to be a true friend, you need a real understanding of a person, of how he or she will react in a host of different contexts. The big difference between online and offline relationships is that in the former you show only what you want to show, often just cataloguing what you like and dislike. No one sees how you really deal with problems or suffer in stressful situations that have real and permanent consequences. By contrast, you cannot so successfully hide from a real friend in a face-to-face situation what you may be truly feeling, especially if your friend is adept at using all the three-dimensional and sensory clues needed for real empathy.

The lack of opportunity online to rehearse social skills might well foretell a decline in deep and meaningful relationships. An important consideration is that a preference for online rather than face-to-face communication could result in greater distrust of people. After all, trust grows from empathy, which in turn is best established through face-to-face communication and body language.

Surely it is when time spent on online relationships replaces time spent on real human interaction that the potential to miss out on deeper intimacy with others is more likely. So we need to think about the impact of Facebook-type relationships on lifestyle in general. Too much social networking can cross the line into interpersonal dysfunction and damage, even demolishing careers and marriages. It can displace time spent on relationship maintenance, and lead to an increased opportunity to communicate with ex-partners or potential future partners, either of which leads to temptation or to jealousy in current relationships. A 2013 study found that high levels of Facebook use were associated with negative relationship outcomes, leading to more cheating, breakups, and divorces. This effect was influenced by how much conflict the couple experienced in relation to Facebook.[29]

Social networking sites now expose users to information to which they wouldn't otherwise have access, such as photos of an ex with a new partner. Thus Facebook can feed the insecure and jealous side of human nature.[30] One friend told me that she left Facebook because it started making her feel paranoid, even though she wasn't an inherently jealous type: "But suddenly there was this information that I could know about my partner, that I didn't want to know, but I couldn't help myself looking for." There are formal studies evaluating and recognizing just this reaction. One investigation was based on an earlier finding that continuing offline contact with a former romantic partner may disrupt emotional recovery from the breakup.[31] Results from 464 participants revealed that those who remained Facebook friends with their ex-partner, compared to those who did not, reported sexual desire and longing for the former partner, combined in a toxic mix with lower personal growth. The researchers concluded: "Overall, these findings suggest that exposure to an ex-partner through Facebook may obstruct the process of healing and moving on from a past relationship."[32]

Of course, this is true in real life as well. It's hard to get over people when you continue to see them routinely. But Facebook makes this unhealthy perseveration so much more accessible and more difficult to resist. Historically, our relationships would be periodically pruned—for example, through the demise of an intimate relationship, a falling-out with a friend, a change in jobs, schools, or residence, or simply losing contact with someone. Now, thanks to Facebook, we can much more readily cart around all that emotional baggage from the past into our present.

Moreover, greater access to others' personal information has led to a culture where snooping on individuals is not only allowed but expected. The Facebook vernacular is "stalking," but social networking site researchers have softened the term to "social surveillance." Regardless of the semantics, the ability to pry freely and anonymously into the lives of others is a serious issue. You only have to look at the popularity of celebrity gossip magazines to realize that humans are inherently nosy. But this tendency can now be amplified through social networking, where interpersonal surveillance is a fairly common practice: 70 percent of college students (the behavior occurs irrespective of gender)[33] reported using Facebook to check up on their romantic partner, with 14 percent reporting doing it at least twice a day.[34] Indeed, an increased level of Facebook use predicts jealousy linked to the social networking site. Investigators argue that this effect may be the result of a feedback loop: Facebook use can expose people to ambiguous information about their partner that they may not otherwise have access to, and this information incites jealousy and further Facebook use.[35]

One law firm that specializes in divorce claimed that almost one in five petitions they processed cited Facebook.[36] Flirty emails and messages found on Facebook pages are increasingly being used as evidence of unreasonable behavior. According to the British legal services firm Divorce-Online, Facebook was implicated in 33 percent of marriage break-ups in 2011, up from 13 percent in 2009. Mark Keenan, managing director of Divorce-Online, commented:

> I had heard from my staff that there were a lot of people saying they had found out things about their partners on Facebook and I decided to see how prevalent it was. I was really surprised to see 20 percent of all the petitions containing references to Facebook. The most common reason seemed to be people having inappropriate sexual chats with people they were not supposed to.[37]

Time spent using technology is time spent away from the real world and real people. It is through seeing others or hearing their voice that we can try to understand how they feel. Too much time focused on the two-dimensional world of social networks may, as we saw earlier, be affecting young people's ability to empathize with others, form meaningful bonds, and ultimately get the best out of their relationships.

In a debate in London in February 2012, I locked horns with Ben

Hammersley, the editor of the magazine *Wired*. The motion was "Facebook is not your friend." It would be unfair to Ben, who has no voice in these pages, to try to summarize the entire interchange of views. However, the reason I raise the occasion here is that in our summing up, Ben conceded that Facebook was indeed your friend because it was "just fun," and obviously it was no substitute for real friendships. In the heat of the emotionally charged moment, I launched into a lengthy riposte, but in retrospect I wish that I had simply acknowledged that Ben had just proved my point. Social networking sites could be as much fun, as insubstantial, and as potentially compulsive as junk food. What seems irrefutable, however, is that such sites are having a significant impact on interpersonal communication and hence relationships. And if that is so, then, as with junk food, there will inevitably be still-wider repercussions for society as a whole.

SOCIAL NETWORKING
AND SOCIETY

The whole point of the term "Mind Change"—as opposed to, say, the potential, sci-fi-sounding term "brain change"—is that it touches on many aspects of how we as individuals think, feel, and interact with one another the longer we live in this unprecedented digital environment. In order to gain the bigger picture, it's important to think not just about the neuroscience underpinning these transformations but also about the psychology, social science, and even philosophy behind them. From the seventeenth century onward, great thinkers such as Thomas Hobbes, John Locke, and Jean-Jacques Rousseau promoted the idea of the social contract, which holds that individuals have consented, either explicitly or tacitly, to surrender some of their individual freedoms or rights for their own ultimate protection and well-being. So let's look now at what impact social networking sites have on the accepted moral values of a society.

Megan Meier was a thirteen-year-old living in Missouri when, in 2006, she started communicating online with a boy named Josh Evans.[1] At first Josh seemed caring, but then became increasingly critical and insulting; he told Megan that she was such a bad person she should kill herself. In fact, "Josh" was the mother of an ex-friend of Megan's. This story not only demonstrates how easy it is to adopt a completely different persona but, much more significant, also illustrates the effects that such bullying can have: Megan did as she was told and hanged herself. Alarmingly, this type of tragic story is becoming increasingly common.

The vulnerability of teenagers to sanitized yet less rich, less multidimensional forms of communication, their age-related propensity to take risks, the twenty-four-hour availability of social networking, and the unedited and unrealistic snapshot social networking sites present of what everyone else is up to are all factors that might prove to be a heady cocktail for some individuals, who could then behave in dysfunctional ways that have eventual implications for society as a whole. In 2012 a survey in the United States, Canada, the United Kingdom, and Australia showed a stark increase in the number of suicides resulting from cyberbullying, with 56 percent of cases occurring in the previous seven years and 44 percent of cases in the previous fifteen months.[2]

Cyberbullying is when someone uses the Internet, a mobile phone, or another device to threaten, harass, tease, or embarrass another person. Various studies report that 20 to 40 percent of young people have been victims of cyberbullying.[3] In a survey of U.S. teenagers in 2011, 33 percent of girls age twelve or thirteen who use social networking sites said that peers' interactions on social networking sites are "mostly unkind," and 20 percent of girls age fourteen to seventeen reported the same thing.[4] Often these bullies will set up a website or form a group on Facebook and get others to join in and make comments about another person. But it isn't fair to blame the Internet for this. Bullying has long cast its dark shadow over the playground and workplace, and it seems deeply ingrained in our psyche.

"Perhaps it is only human nature to inflict suffering on anything that will endure suffering, whether by reason of its genuine humility, or indifference, or sheer helplessness." This suggestion came from Honoré de Balzac in his 1835 novel *Le Père Goriot*.[5] It's even been hazarded that bullying has evolutionary value as a stabilizing factor in the shifting struggles for hierarchical status in primate colonies.[6] But while bullies have been a blemish on society ever since Flashman, for example, strutted his stuff in *Tom Brown's Schooldays,* the vehicle for them to express their nasty predispositions has changed. Now that the Internet and social networking have removed most constraints on accountability, it is possible that this technology could result in behaviors and situations that previously wouldn't have been possible.

Some will argue that the effects of the digital culture on cyber bullying is a non-issue, because the medium is irrelevant. For example, Dan Olweus, who runs a bullying prevention program at Clemson University, found in a large sample of younger teens that there was a high degree of overlap between those who bully in traditional ways and those who engage in cyberbullying. However, 12 percent of new victims or bullies in the U.S. sample were cyberbullying only, and had not been victims or bullies in the traditional way. Olweus argues this is a "very small percentage," and he goes on to say,

> These results suggest that the new electronic media have actually created few "new" victims and bullies. To be cyberbullied or to cyberbully other students seems to a large extent to be part of a general pattern of bullying, where use of the electronic media is only one possible form and, in addition, a form with a quite low prevalence.[7]

However, the 12 percent of young teenagers who participate in bullying or are victims of it is hardly a "very small percentage." Moreover, we need to ask not only whether the Internet encourages this behavior but also, and more important, whether cyberbullying can affect a victim more seriously than traditional bullying. After all, the scale of the audience that can witness the bullying is now much greater than would have been the case with traditional bullying, and evidence of it can exist permanently on the Internet. A recent study found that both cyberbullies and their victims were significantly more likely to internalize problems, as evidenced by depressive symptoms and suicidal behavior, compared to those involved in traditional bullying. So the medium can affect both victim and bully much more seriously.[8]

Experts have argued that the Internet creates a unique world that adds extra "disengagement" from immoral actions.[9] The process of moral disengagement describes how an individual is able to deactivate internal moral controls that otherwise inhibit his or her behavior.[10] This disengagement may be a prerequisite for cyberbullying: visual cues such as the victim's distress are absent, while the distance created by a screen suppresses any feelings of guilt and shame. Furthermore, because young people associate the use of technology with online games, chatting with friends, and exchanging photos, cyberbullying is often closely connected with other means of entertainment.[11] This finding is in line with research showing that cyberbullies have less remorse, which may be due in part to the lack of direct contact between the bully and the victim. Two investigators, Sonja Perren and Eveline Gutzwiller-Helfenfinger at the University of Zurich, found no relationship between moral disengagement and cyberbullying.[12] This suggests that the screen may dehumanize victims to such an extent that bullies do not even need to suppress any moral values, therefore they don't need to first disengage, to harm others online.

Diffusion and dilution of responsibility are other drivers of cyberbullying behavior.[13] Just as bullying in the presence of a gang permits the responsibility for the act to be diluted, cyberbullying often takes place within a virtual crowd. The Internet offers the anonymity of a mob, and thus the opportunity to behave in a more shameful way than one would in person. Dr. Graham Barnfield, a media researcher and lecturer at the University of East London, told the British TV program *Tonight with Trevor McDonald* that "happy slapping"—when bullies film bullying on their phones and upload it to the Internet—can be seen by the "slappers" as a short-cut to "fame and notoriety." This is an example of a completely new kind of mentality made possible only by the Internet.

There's another phenomenon that, like bullying, seems also to bring out the worst in human nature and, like happy slapping, could only really happen on the Internet. Trolling is prevalent in chat rooms, Twitter streams, and blogs. The concept of "trolling" generally describes someone who adopts an offensive or controversial stance in order to annoy others or to provoke an emotional response.[14] Mature and seasoned Internet users may take trolls' comments with the appropriate pinch of salt, especially if they are more witty than spiteful, but more sensitive users or impressionable younger victims may take offense or have their self-esteem and confidence demolished.

It could be, of course, that a certain unpleasant type of person naturally enjoys causing offense no matter what and might have found in the Internet merely another outlet. But it is hard to imagine how trolls might truly express themselves face-to-face with their victim in the real world. For example, in one terrible case Internet trolls contacted a bereaved mother, pretending to be her dead young daughter getting in touch from hell.[15] Extreme though this example may be, it illustrates how an environment of widespread global access, diminished responsibility, and anonymity, combined with a lack of experience of interpersonal relationships, have pushed trolling to new heights, or more accurately lows.

John Newton, head of a school in Devon, wrote of his concerns in a national British newspaper, the *Daily Telegraph,* suggesting that social networking websites pose a serious threat because they blur the lines between gossip and fact before schoolchildren learn to appreciate the difference.[16] Newton has warned that social networking sites are "a far more powerful weapon in the hands of our children than we appreciate." Of Facebook in particular he asks:

Is it a meaningful social hive generating goodwill and reuniting old friends, or is it a gossips' paradise infesting the world with innuendo, half truth and insult? If they flippantly post comments on-line, young people especially may not realize the irreversible consequences to someone's reputation. . . . They have not necessarily understood what constitutes gossip, nor appreciate the Exocet quality of a hurtful word; half-formed opinion is all that counts.

This picture of a more malicious and less moral society driven by social networking sites may not hold for all societies because of differences in the way cultures use such sites. One investigation compared social networking site use in a collectivistic culture, China, and in an individualistic one, the United States.[17] More than four hundred college student participants were recruited from a southwestern university in China, and a comparable number from a midwestern university in the United States. The participants completed a survey about their use of social networking, including time spent on it, its importance to them, and their motives for using it. There were clear cultural differences. U.S. users spent more time on social networking sites, considered them to be more important, and had more virtual friends than their Chinese counterparts did. These findings suggest that in collectivistic cultures the importance of family and friends may be partly responsible for Chinese users' weaker ties to social networking. In contrast, individualistic cultures may offer less support for close and enduring friendships, resulting in greater use of Facebook and the like. Given the evidence so far that social networking promotes an individualistic focus, it's surely unsurprising that the Western world seems to use social networking in ways not paralleled in Eastern cultures.

Despite accumulating evidence of the dark side of social networking,[18] the potential to spread information at breakneck speed in countries where information may be repressed or controlled is a vital tool. Facebook and Twitter use among activists played a key role in the Arab Spring uprisings in 2011.[19] Moreover, social networking may be an effective means of raising global consciousness among users—for example, to encourage young people in the United States to vote and to impart awareness of humanitarian plights. In turn, large sums of money can and have been raised by crowdfunding, the collective effort of individuals who network and pool their money via the Internet to support efforts initiated by others in support of a wide variety of activities, from disaster relief to start-up companies.

What effect is this "clicktivism" having? For example, did liking or sharing the *Kony 2012* video to stop war criminal Joseph Kony change a user for the better? The rate at which individuals participated in the Cover the Night activism proposed by *Kony 2012* was significantly lower than would have been predicted, given the immense popularity of the video. An outstanding issue in clicktivism is how to translate what's on the screen into real-world actions.[20] Social networking sites can provide us with large quantities of information about world issues, and clicktivism requires next to no effort while making users feel good. Others have termed this kind of passive, easy concern "slacktivism." Indeed, given the research we've discussed showing that the screen can sanitize interpersonal communication and dehumanize individuals, viewing humanitarian crises through a social networking site may have less impact than if a user was exposed to the situation offline. Clicktivism could well reduce the incentive to make a credible impact on humanitarian issues, because a user feels that liking and sharing a cause has been enough.

Drawing on interviews with teens and young adults, one study explored the extent to which young people's approach to online life included moral or ethical considerations.[21] The data revealed that individualistic thinking was the primary focus when making decisions online; community-focused thinking was least prevalent. Moreover, nearly all individuals in the study could identify at least one instance in which they had trivialized the moral elements of online activities, indicating that individuals have a "greater tolerance for unethical conduct online." Perhaps we are indeed in danger of forgetting John Donne's famous lines:

> *Any man's death diminishes me,*
> *Because I am involved in mankind,*
> *And therefore never send to know for whom the bell tolls;*
> *It tolls for thee.*[22]

Facebook, Twitter, and similar sites deliver the promise of being constantly connected, wanted, admired, even loved. They have brought into our society interpretations of identity and relationships that challenge current values and morality, in a way that would have been inconceivable even just a decade ago.

THE *SOMETHING* ABOUT VIDEOGAMES

There's no point in having fun. But surely that *is* the point: to focus entirely on an activity in the here-and-now present, an end in and of itself. Yet there may be more to it than that. Since the dawn of time human societies have appreciated the importance of fun, often in culturally institutionalized events such as parties and feasts. Wine, women, and song and their more modern reincarnations, sex, drugs, and rock and roll, free us up to live for the moment, to have our raw senses directly stimulated, with no time for abstract thoughts and self-conscious introspection. And all this fun could actually have serious, evolutionary value. Immersion in a sensation-soaked present would favor participating in the material, immediate joys of reproduction and nutrition that are essential to survival.

It needn't stop there. Talents rehearsed across the card table or playing charades on a rainy winter evening translate directly to becoming proficient in the interpretation of body language, in knowing how to employ eye contact, and in learning to empathize generally with thought processes and emotions, as well as in developing important cognitive skills such as reasoning and memory. Playing with dolls anticipates caring for babies, while all types of sports develop the teamwork, physical health, and competitive skills that in the primeval savanna would have ensured the survival of the fittest. Yet videogames could, for the first time, be dissociating fun from any of the survival-value requisites that traditional games have met. Rather than serving as a twenty-first-century answer to age-old needs, the gaming experience could be an end in itself rather than a means for thriving in the real world.

The advent of the smartphone has further transformed the experience: 36 percent of American gamers access games on their smartphones,[1] and it seems that this trend will increase in the future as phones become increasingly personalized. These vast technological advances make videogames richer and more diverse experiences and have contributed to their soaring popularity. Interestingly enough—and contrary to earlier trends—games are rapidly becoming popular with older generations. The average age of a gamer is now estimated at around thirty, with 45 percent of gamers being women.[2] Nonetheless, videogames readily offer something that appeals to people of all ages, backgrounds, and cultures, which the real world and traditional games only rarely do.

Nicole Lazzaro is the founder and president of XEODesign, "the world's first player experience design consulting company," and a leading researcher "on emotion and the fun of games." An authority on emotions and videogames, Lazzaro identified four different types of fun, with the best-selling videogames offering at least three out of four on the list. *Hard fun* gives you challenge combined with the promise of eventual mastery; *easy fun* provides the sheer joy of the gaming experience itself. *Serious fun* enlivens your otherwise dull tasks, while *people fun* is the inevitable result of hanging out with your friends.[3] Real life rarely provides more than one or two of these opportunities at the same time and certainly not on demand, whereas videogames are meticulously designed to do just that.

However, not all games are created equal. They can vary not

only in their platform (e.g., PC, gaming console, mobile phone) but also in their mode (e.g., single player, multiplayer, offline, online). First-person shooter games remain popular in both online and offline modes; one of the most sought-after titles, the latest version of the Call of Duty series, sold more than twenty-seven million units within the first six months of release.[4] While incentives to play differ according to gender, age, personality attributes, and mood of the gamer, a few common elements in the appeal of videogames rank consistently high as factors. Players most frequently cite opportunities "to achieve," "to escape," and "to socialize" as reasons for entering these unreal worlds.[5]

Videogames have existed for more than half a century, but only in the past two decades have they become endless collaborative online experiences, often with thousands of other human players interacting simultaneously. Known as massively multiplayer online role-playing games (MMORPGs), they focus on the progression of the gamer-controlled character, an avatar, in a fantasy world. Unlike typical first-person shooter games, MMORPG players have full control over the physical features, development, and attributes of their avatar. Avatar progression is realized through combat, exploration, item acquisition, skill development, socialization, and narrative. The MMORPG world in which this action unfolds is much larger in scope than the worlds of first-person shooter games, with such large numbers of players able to interact in the same virtual world simultaneously. Furthermore, this global game is persistent, in that, regardless of whether or not a gamer is logged-in, the world continues to turn in the cybersphere, updating and evolving. In contrast, first-person shooter games are typically made up of purely "instance" scenarios, in which the plot only exists for the duration of the game and can be restarted at the original point an infinite number of times.

This distinction is important. In a review of the current findings on gaming, authors Daria Kuss and Mark Griffiths conclude that, given the endless possibilities of the new worlds of MMORPGs, their social nature, and the possibility for the gamer to develop an attachment to their avatar, MMORPGs are the most addictive type of videogame.[6] A friend whose son had become hooked on videogames to the exclusion of much else and who can himself see the appeal of, and vulnerability to, gaming, tried to explain: "The games are designed to pull the player in, to ensure that each level is rewarded by the next level, that play never naturally stops, and that if you take a break you either suffer in the game play or you feel desolate as a result of the lack of exciting and rewarding game play."

This personal attachment and emotional investment in an alternative gaming self adds to the lure. Experiences are designed to provide excitement and meaning as a means of manipulating behavior. As the designer of Gamasutra, a website founded in 1997 that focuses on all aspects of videogame development, John Hopson has been able to analyze the attraction in terms of established behavioral theory, where conditioning is used to teach both humans and animals new information and behaviors. For example, rats can be controlled by being rewarded with food pellets or punished with shocks in accordance with a simple behavior such as pressing a bar. Hopson has described how, like a rat, a gamer can be manipulated into continuing to play when a reward is given only under specific circumstances. In certain schedules, not only are the rats avoiding the unpleasant, but they are also hooked by the uncertainty of not knowing when a reward will come; they just know that one will eventually come along.[7] For human gamers, there may be a reward after a number of actions have been completed (fixed ratio schedule) or, alternatively, after a specific number of actions, with the number changing every time (variable ratio schedule). Then there are what are known as chain schedules, with multiple stages to achieving the goal, where the player needs to respond quickly. Games are thus excellent vehicles for manipulating brain processing at a very basic level.

Back in 2002, social scientist Nick Yee conducted seminal research on nearly four thousand players to gain more insight into gaming behavior.[8] He found that well over half of all the gamers fessed up to playing MMORPGs continuously for ten hours or more in a single sitting, and over 15 percent reported feeling anxious, irritable, or angry if they were unable to play. Nearly 30 percent admitted they continued to play even when they became frustrated or upset or had stopped enjoying the game, while 18 percent claimed that gaming had caused them academic, health, financial, or relationship problems.

Many of us have our own stories to tell. One father told me:

> Having had a son who lost a year of university through playing World of Warcraft, I nevertheless believe that the fact that he has moved on from that game and is now in a successful career (for the moment touch wood!) does not mean he is free of the gaming addiction. He is not, and I doubt if he ever will be.

This father, a friend of mine, was almost in tears the first time he told me of his son's plight, and for quite a few months it was the sole topic of conversation whenever I saw him. He and the boy's mother felt guilty and out of their depth: when and how had this happened?

Any behavior might have addictive qualities—that is, it may be characterized by a compulsion to engage over and over in an action until that action has a serious and persistent negative effect on an individual's physical, mental, social, and/or financial well-being. As of May 2013, "Internet use disorder" has been included in the fifth edition of the *Diagnostic and Statistical Manual of Mental Disorders* (*DSM–5*) as a condition "recommended for further study" within Section III. This move effectively postpones full recognition and inclusion of the disorder until uniform criteria needed for a robust psychiatric diagnosis can be agreed on.[9] For more than a decade, however, numerous studies have produced evidence that excessive use of the Internet, and of related features such as gaming, may be considered a behavioral addiction comparable to pathological gambling.[10] But herein lies a snag: not all Internet activity involves gaming, and vice versa.[11]

Then again, when researchers study specific features of excessive Internet use, online gaming is the most frequently explored. Despite the multiple ways we might conceptualize excessive gaming and measure the behavior, two symptoms appear consistently: significant problems as a result of overuse of games, and an inability to control use. Some hallmarks of gaming addiction include lying about how much time is spent gaming; intense feelings of pleasure or guilt; spending more and more time gaming to get the same enjoyment; withdrawing from friends, family, or spouse; experiencing feelings of anger, depression, moodiness, anxiety, or restlessness when not gaming; spending significant sums of money on online services, computer upgrades, or gaming systems, and thinking obsessively about gaming even when doing other things.[12]

Some argue that these screen-based experiences are just a medium through which an addictive activity is accessed.[13] In other words, the gambler who drives every day to the casino is addicted to gambling, not to driving. Similarly, the person who uses the Internet to gamble is addicted to gambling, not to the Internet. However, while online gambling has a real-world alternative option, gaming, by definition, does not. Unlike gambling, it's a phenomenon specific to digital technology. Hence any abnormal behavior associated with videogames cannot be divorced from the medium of the screen and the unique experience it gives. So while much Internet activity might encompass gaming, we need to remember that an addiction to gaming will be an addiction to gaming, and not to anything else.

Statistics on Internet addiction vary widely across cultures and depend on the form of evaluation used.[14] However, the numbers for gaming addiction specifically seem to be much more consistent. Drawing on a U.S. sample, Douglas Gentile found that 8 percent of gamers between eight and eighteen years old were classified as addicted,[15] while another recent review gives an estimate of 2 to 12 percent.[16] Moreover, a ballpark figure of just under 10 percent seems to be consistent across continents: in a two-year longitudinal study performed with a general elementary and secondary school population in Singapore, including some three thousand children in third grade, the prevalence of "pathological gaming" was similar to that in other countries, 9 percent.[17] But leaving aside questions of definition, conflation with other Internet activities, statistics, and the additional appeal of online interaction, can we say with any certainty that gaming is addictive?

Aviv Weinstein at the Hadassah Medical Organization in Jerusalem believes that the craving for online gaming and the craving for substance dependence could well share the same neurobiological mechanism.[18] Weinstein argues that teenagers may play longer, prioritize thinking about games over other important matters, game to escape emotional problems, have difficulties with academic work and socializing, and conceal gaming activities from their family. Individuals with this behavioral pattern who then experience intolerable irritability when they stop gaming are displaying the classic hallmarks of an obsession, even an addiction. But can the behavioral commonalities of traditional addiction and intense gaming be linked to the same single brain state?

In one particular brain-imaging study, healthy control subjects displayed a reduction in the number of molecular targets (receptors) for the neurotransmitter dopamine in a key brain region (the ventral striatum) after playing a motorbike riding computer game. In striking contrast, former chronic Ecstasy users showed no change in the status of their receptors after playing this game.[19] For the nonaddicts experiencing the thrill of gaming, there was a surge in the release of dopamine that "desensitized" its receptors (remember the handshake analogy in Chapter 7 and how a hand would become numb when pressed for too hard or too long). But the brains of the Ecstasy addicts told a different story. Here chronic use of the drug had accustomed the brain to vast amounts of dopamine. The videogames added no excitement because they worked via the same common mechanism. *It seems that as far as the brain was concerned, taking Ecstasy and gaming were comparable experiences.*

Another way of showing that gaming releases high levels of dopamine in the brain is through looking at changes in the actual size of brain structures. Do you recall how the hippocampus was bigger in London taxi drivers because they constantly relied on their working memory while driving? It seems the same principle might apply to gamers and their dopamine systems. In young gamers, brain imaging shows an enlargement of the area of the brain (the ventral striatum)[20] where the neurotransmitter dopamine is released.[21] Interestingly enough, a similar feature is also characteristic of the brains of pathological gamblers, who suffer from another behavioral addiction.[22] So it seems that whether we're talking about addiction to drugs, gambling, or videogames, all three conditions are linked to excessive dopamine release in the ventral striatum.

The next question is whether individuals with brains that happen to have an enlarged ventral striatum are predisposed to gaming, or whether excessive gaming has literally left its mark on the brain. This presents a tricky chicken-and-egg conundrum that in general bedevils brain research: does an unusual brain feature cause an unusual behavior, or does an unusual behavior change the brain, thanks to its plasticity?

Let's start with the chicken: that gaming, as with all life experiences, leaves its mark on the impressionable, plastic brain. The work of Simone Kühn's team at Ghent University in Belgium would suggest that this is the case. They showed that gaming could be linked to a larger striatum, reflecting adaptive neural plasticity through the sustained release of dopamine.[23] In other words, the more time spent playing, the more pronounced the expansion of the striatum. This suggests the former caused the latter.

Then there's the egg: the idea that a preexisting brain state might predispose individuals to compulsive gaming. Kirk Erickson at the University of Illinois found a correlation between the volume of a key brain area, the dorsal striatum, and later training success in a videogame.[24] Erickson has also described a link, again seen with imaging, between activation of the striatum before training and subsequent later skill acquisition during gaming. These findings highlight the importance of the striatum, a rich source of dopamine, and how this might be consistent with the idea that some brains are more susceptible to the lure of games. Individuals who happen to have a larger striatum might experience gaming as more rewarding in the first place. This neurological setup, in turn, could facilitate skill acquisition and lead to further rewards from playing.

So which came first, the chicken or the egg? Did a strong and sustained experience shape the brain, or was a certain type of brain already predisposed to respond readily to that experience? An important clue to the right answer is the anatomical makeup of the striatum itself. This structure can be divided into two parts, an upper (dorsal) zone and a lower (ventral) one. It turns out that the latter releases more dopamine than the former.[25] So it may not be surprising that the two regions have been associated with different kinds of functions: the dorsal striatum coordinates sensorimotor functions for *attaining* a goal, while dopamine released from the ventral part *enhances the impact* of the actual reward that then ensues.[26] So one way of resolving the chicken-and-egg problem might be to say that a brain predisposed to effective sensorimotor coordination, with an active dorsal striatum, will have a predisposition for gaming, while it is the games themselves that change the way the ventral striatum reacts to reward. Yet neuroscience is rarely so simplistically cut and dried, and certainly research in this area is still in its infancy.

In any case, the chicken scenario, where obsessive gaming directly impacts brain states, is not necessarily mutually exclusive with the egg scenario of a particularly predisposed type of brain. The most significant issue here is the contribution of dopamine.

Neuroscientists at Hammersmith Hospital in London have shown that playing videogames directly results in release of this neurotransmitter.[27] However, just as it is impossible to establish a causal link between, say, the known biochemical actions of Prozac in enhancing serotonin availability and the alleviation of depression, so the translation of the objectively observed release of dopamine in the brain into the subjective effects of feeling wildly happy is very difficult to conceptualize.

There is nothing magical locked inside the dopamine molecule. Rather, it is the particular site in the brain, together with the environmental context in which it operates, that determines the final net effect. Suffice it to say that raised levels of dopamine are consistently linked to various brain states relating to arousal, reward, and addiction. Moreover, the idea that the chicken and egg are not exclusive but may actually be reinforcing each other would be a great example of how the brain and the environment are in an intense and continuous two-way dialogue.

So why do some people and not others become addicted to videogames? Perhaps an individual's capacity for excitement could be pivotal. Since dopamine is linked to high arousal, as can be seen with the drug amphetamine (that causes the release of dopamine in the brain), this idea is quite logical. One investigation found different patterns of arousal in different types of gamers. Those who played excessive amounts of first-person shooter videogames had significantly higher levels of arousal during gaming, which dropped off immediately after gaming.[28] By contrast, gamers who did not play excessively stayed "high" even after gaming had finished. MMORPG players who gamed excessively displayed significant *decreases* in physiological arousal, which rose again immediately after gaming. Meanwhile, MMORPG players who didn't play excessively showed normal increases in arousal during gaming and then reached a plateau after their session.

These differences in arousal for gamers of different genres are comparable to those reported in the scientific literature on pathological gambling. There are the thrill-seeking, impulsive addicts who take stimulating substances or engage in high-risk behaviors; by contrast, there are the escapists, the often depressed addicts who are not seeking high arousal.

So for the second type of player, the kind who experiences low arousal, the time spent in MMORPGs and the meaningless nature of the activity could have long-term implications for mental state. Of course, once again the chicken-and-egg complication applies: these disturbances in arousal regulation could be either a cause of gaming addiction or a consequence of it. Nonetheless, the discovery that the activity physiologically affects excessive gamers differently than gamers who don't play to excess is an important consideration to bear in mind.

Yet at the end of the day, what finally determines an individual's level of arousal and whether he or she becomes addicted to one or another type of videogame? This is impossible to answer, as is the question why certain individuals and not others are predisposed to being kind or shy or funny. There may be some extremely indirect, genetically based predispositions. For example, a possible inherited vulnerability to videogame addiction has been reported in studies on the genes encoding for a subtype of dopamine receptor.[29] This in turn might influence the effects of dopamine released in the brain, but even here the causal link is impossible to establish. Remember that it is very unlikely that there is just a single gene for any complex cognitive trait.

It is impossible to tease out a cause-and-effect sequence of events as the brain interacts with the environment, and therefore it is hard to predict with any accuracy whether someone will become addicted to videogames. There are likely to be cumulative effects from risk factors such as low socioeconomic status, parental depression, parental criminality, domestic violence, and parental alcohol and substance abuse, which may be offset to a greater or lesser degree by protective factors. But in the case of my middle-class friend and his expensively educated son, none of those risk factors applied.

A more plausible view would be that what goes on in the brains of those addicted to videogames is not *qualitatively* different but rather *quantitatively* different from what happens in the brains of those who are less obsessive. Why else would videogames be rewarding, by definition, to every single person who plays them? It seems that gaming can induce enough dopamine production to keep the user feeling good, but not enough to completely desensitize the user to its effect. However, the lure of gaming operates not just at the mechanistic biochemical level of brain dopamine but also at the more cognitive one of social relations.

A compulsion for gaming must involve not just the internal machinations of the brain *but also the brain interacting in an ongoing two-way relationship with the screen.* The very nature of this screen environment is crucial in keeping the individual playing. Games are noisy and have visually rich scenes, just like an action movie. In addition, however, they are immersive, offering not just strong sensory stimulation but "flow," or the capacity for a gamer to lose him- or herself in the game world and become utterly involved.

"Playing [World of Warcraft] makes me feel godlike. . . . I have ultimate control and can do what I want with few real repercussions. The real world makes me feel impotent . . . a computer malfunction, a sobbing child, a suddenly dead cell phone battery—the littlest hitch in daily living feels profoundly disempowering."[30] So claimed English professor Ryan Van Cleave, recalling a time when he was playing videogames for some sixty hours a week. Note that Ryan never even mentions "having fun" and that his mindset is signifying something much more profound. World of Warcraft was, for him, a refuge from a real world where he felt inadequate.

Olivia Metcalf from the Australian National University, who has studied the psychology of excessive gaming, makes the distinction that the appeal may not be a positive consequence of the videogames themselves, but rather that the games provide an escape from a purposeless, directionless real life:

Perhaps videogames are more than just escapist fun; they give disillusioned youth the chance to fulfill those needs so intrinsic to being human: competency, purpose, success, achievement, and so on. Research indeed suggests that these are some of the motivations to play videogames: the chance to be outstanding when in real life we are probably average.[31]

The "human" challenge of interacting with other players projected through cyberspace perhaps creates an even greater compulsion for many gamers. As such, online videogames have a higher addictive potential than offline games. Specifically, MMORPGs are thought to have a number of unique characteristics that give them higher addictive potential then other genres. Dr. Daniel King, a senior research associate in the School of Psychology at the University of Adelaide, has conducted an extensive review of worldwide research into "pathological" or harmful videogame playing behavior and found that social interaction is important in the development of excessive gaming. Games with avatars that players can control and identify with are associated with higher addictive potential. These qualities account for why excessive gaming is most commonly seen in MMORPGs. King also found that gamers who play excessively value the achievements gained through gaming, and he proposes that the reward structure built into the game influences the development of excessive gaming.[32]

While MMORPGs have intricate reward systems built into the games, with gamers constantly trying to reach the next level, it is the social interaction with other players that appears to be the real extra hook. Perhaps the appeal is that the player is now not just playing a game but playing out an idealized life that is simultaneously exciting and safe, both physically and mentally. The real world is messy and ambiguous: real people are never either wholly good or wholly bad, they always have inner thoughts or secrets, and actions always have consequences, however indirect, with long-term repercussions that cannot be reversed. Furthermore, in the real world feedback, especially positive feedback on your achievements, is all too hard to come by. And as for life goals, they are for most of us far from clear and usually too complex and provisional to define categorically. According to Nicole Lazzaro (who clearly likes lists of four items), videogames remove much of what is difficult and confusing about real life, since games (1) simplify the world, (2) suspend consequences, (3) amplify feedback, and (4) set clear goals.[33] This inventory may add up to the crucial *something* about videogames that makes them such a compelling escape from the uncertainty and complexity of the real world.

More generally, sometimes the real world is not the best place to be. In some cases games can provide calming routines for people who are unable to cope with the frenetic uncertainty of life beyond the screen. Unlike traditional real-world games, videogames offer a total escape from the dull, difficult world to one that is not just more exciting and sensational (i.e., appealing to the senses) but where there are reassuringly definite and predictable outcomes in which the player can participate as a better self. Research shows that when people are unhappy or dissatisfied with their lives, they create avatars that are very different from themselves.[34] Happy people create avatars just like themselves. Game enjoyment is inversely related to the avatar-person similarity; that is, individuals who are unhappy and create an avatar very different from themselves end up enjoying the game world more. They are literally exploring a new identity for themselves in this game world, choosing an avatar that is better, faster, fitter, stronger, thinner, taller, prettier, or smarter than they are or probably ever can be. Perhaps that's the crux of why videogames may be so pernicious. For most people they remain a form of entertainment, but they open up an entire new world where everything is better than real life, which is particularly appealing to the psychologically vulnerable. And that could be pretty much all of us.

We saw earlier that identity is not just about having a fully fledged mind, which enables you to make sense of the world, but also involves a crucial next step: the reaction you would show, as a result of how you interact with the world, in a specific context at a specific time. In games, however, instead of *your* family, *your* soccer team, *your* choir, or *your* colleagues, the all-important momentary context that is accumulated through the cause-and-effect chain of a unique life story will now be more standardized. Gamers become extremely emotionally dependent on their avatars. They are as attached to their avatars and their teams as someone in the real world may be attached to their real-world relationships. In these instances, the momentary context has shifted online into an artificial world. And what if so much of your life story isn't a story at all, not a sequence of events but, as is the case with first-person shooter games, an atomized, fragmented set of experiences that have no consequences in the real world? In either case, you might start to feel uncertain about who you actually are.

This insecurity could be compounded by a sneaky feeling that what you are doing lacks any real significance or meaning. Meaning, as I've suggested already, can be interpreted through the prism of neuroscience as making connections, of seeing one thing in terms of another. And this can also apply to causal connections over time. This connectivity, as we've seen, has a corresponding parallel in the physical brain as neuronal connections are forged and strengthened through the remarkable plasticity of the human brain. So just as a wedding ring, a simple gold object, can acquire a complex meaning or significance by the associations that develop around it, so significance can be attached to a cause-and-effect linkage.

If you climb a tree and then fall out and break your leg, an injury that takes time to heal, the whole episode will be a meaningful one, not least because it is irreversible. Of course, your leg may well become fully healthy once more, but the event of the breaking itself cannot be airbrushed out. It has enduring consequences in changing forever, one way or another, your view of tree climbing. By contrast, if you drop a bit of paper on the floor and pick it up again immediately, perhaps that's the nearest you'll get to turning back the clock in the real world. It would also have been a pretty meaningless thing to do.

So meaning can be directly related to consequences over time. But if gaming must have, according to Lazzaro, no consequences, it could be regarded as a meaningless way of spending time. And if someone is going to spend all his spare time engaged in a meaningless activity, it may jeopardize in the long term any significance he eventually attaches not just to that activity but, most important, to himself. Yet for the player untroubled by such possible long-term existential concerns, there is the opportunity to simplify and improve instantly on the immediate environment and how he feels within it. The *something* about videogames is that they create a world where you feel good not only because you're having fun but also because you're shutting out the kinds of experiences that would normally make you feel sad, anxious, or worthless. You enter a world designed to cater to your psychological needs; therefore there will be a complex and wide range of effects on how you think and feel over the longer term. "What we do know," concludes Daphne Bavalier, an expert in this field at the University of Rochester, "is that, in technology, we have a set of tools that has the capability to drastically modify human behavior," inevitably by modifying the brain.[35] What is needed, she feels, is a way to ensure that technology is specially designed in order to achieve desired outcomes. But this might be easier said than done.

We've seen here that videogames can be affecting mental processes in a complex and diverse range of ways. There are a host of different questions that have to be unpacked separately. For example, if the gaming reward schedules are locked into a fast iteration of stimulus and response, then what effect might prolonged gaming have on attention? Moreover, given that violent games account for 50 percent or more of all videogame sales, will playing these games increase aggressive behavior in the real world?[36] Finally, if, as we've seen, there is no permanent meaning in the escapist gaming world because actions don't have enduring consequences, will it result in people becoming more generally reckless in real life? Let's explore each of these issues in turn.

VIDEOGAMES AND ATTENTION

"The sounds of silence are a dim recollection now, like mystery, privacy and paying attention to one thing or one person at a time," writes *New York Times* columnist Maureen Dowd, looking wistfully back to another era.[1] Perhaps we shouldn't be too surprised that if nowadays we end up engaged for hours in activities bombarding us with fast-paced stimuli, then our exquisitely adaptable human brain will obligingly adapt to that environment, an environment that does not require sustained attention. And the more stimulation flooding in, the shorter the attention span that can be allocated to each input. So could videogames, given their fast-paced and vivid content, be affecting attention in a way that is unprecedented and unique compared to all the usual, more muted distractions of real life?

Before we can even think about answering this question, we need to sort out the common and understandable complaint that the Internet in general, and gaming in particular, is to blame for a range of problems that also might be justifiably generalized to human nature, the modern world as a whole, or at least any screen-based technology, such as the good old TV. Such critics have a fair point. For example, at the Seattle Children's Hospital, Dimitri Christakis examined more than a thousand children at one year of age and a similar number at age three.[2] He found that 10 percent of the children sampled had attentional problems at seven years of age that were linked to the number of hours of television that had been viewed per day between the ages of one and three. So while shortening the attention span is obviously not a good thing, gaming can't have any *additional* impact compared to other, older screen-based experiences . . . or can it?[3]

Edward Swing and his team at Iowa State University have conducted the first long-term study on the specific effects of videogame use by elementary school children.[4] The project involved 1,323 children between six and twelve years old who, together with their parents, recorded their television and videogame exposure at four points over a thirteen-month period. Teachers measured attention problems by reporting difficulties the participants had staying on task and paying attention, and whether a child often interrupted another child's work. It turned out that those who had more than two hours of screen time (television and videogames combined) per day were more likely to be above the norm in showing attention problems. However, the results also revealed that playing games was linked specifically to a greater risk of developing attentional problems, and that it was in fact a more robust predictor than television viewing. Even after allowing for the effect of TV exposure, as well as any earlier attention problems the child already had, the amount of time spent playing videogames by each child accurately predicted increases in problems with attention just over a year later.[5] So gaming would seem to have a specific detrimental effect.

Subsequent research has investigated in more detail the relationships between gaming and attentional problems and reached similar conclusions. At Iowa State University, Douglas Gentile and his team followed up with a sample of more than three thousand children and adolescents tracked over three years.[6] Children who spent more time gaming had more attention problems, even when earlier attention problems, sex, age, race, and socioeconomic status were statistically controlled for. Interestingly enough, children who were more impulsive or had more attention problems subsequently spent more time playing videogames, indicating a possible bidirectional effect of gaming on attention problems: the one enhances the other, and vice versa.

These investigations provide the strongest evidence to date that the association between videogame play and attention problems is not coincidental but causal. This possible interrelationship has potentially interesting implications for Mind Change. It demonstrates clearly how the brain and the environment are in such constant dialogue with each other that it's often hard to tease out the chicken and the egg, as we've seen already. Someone who is impulsive and readily distracted might find in videogames the perfect vehicle for his or her disposition, while habitually spending time in a world mandating quick reactions and instant feedback will guarantee that the brain adapts to that fast-paced environment.

Modern videogames, with their visually rich and fast-paced play, are likely to place significant visuo-spatial and cognitive demands on a gamer, and these demands will in turn leave their mark via the plasticity of their brain and hence on the individual's subsequent behavior—but not necessarily with negative consequences. Research shows that gamers make excellent drone pilots, and even outperform real pilots on certain tasks.[7] In the same spirit, scientists at the Duke School of Medicine have investigated just how effectively skilled gamers might eventually become highly proficient drone pilots, compared to their student colleagues who didn't play action games.[8] Greg Appelbaum, an assistant professor of psychiatry, set the subjects a visual memory task to see how efficiently they could recall information they had just seen for the first time. The experienced gamers beat their rookie counterparts, proving that they could respond to visual stimuli much more quickly. This draws upon the skill needed in first-person shooter games, where gamers need to decide what to "blast" every second. "Gamers see the world differently. They are able to extract more information from a visual scene," Appelbaum concludes. "They need less information to arrive at a probabilistic conclusion, and they do it faster."[9]

Some researchers have suggested that it is in fact the *motivations* of gamers that can create differences between gamers and nongamers, rather than superior visuo-spatial skills.[10] Think about it: gaming enthusiasts spend their free time using computers for the enjoyment and competition of game tasks, whereas nongamers recruited into different studies obviously will not have a preference for such activities if other options are available. Thus perhaps it's simply that gamers have a certain mindset leading them to be more competitive, to enjoy computer tasks, or to be more incentivized to do well in the scenarios that result in the visuo-spatial improvements.

A whole host of different processes and functions, such as vision and motor control, appear to be enhanced by regular gaming.[11] Compared to nonplayers, seasoned action gamers have demonstrably better hand-eye coordination and visuo-motor skills, such as resistance to distraction, sensitivity to information in the peripheral vision, and an ability to count briefly presented objects. With the development of the PlayStation Move, Kinect, and Wii, videogames can also lay persuasive claims to developing motor skills by encouraging full-body movement.

One of the key studies showing the beneficial effects of gaming took place as long ago as 2003, when Shawn Green and Daphne Bavelier at the University of Rochester investigated the impact of action videogame playing on vision. They were interested in whether learning could improve performance in different tasks other than those on which the training was focused. Initial experiments confirmed the expected improvements: in different aspects of visual attention (the ability to focus on one part of the visual field), the habitual videogame players outperformed the rookies. Most significant, however, was that in a final experiment the non-players who had been subsequently trained on an action video-game showed a marked improvement that transferred to skills well beyond the training task. Green and Bavelier concluded, "There-fore, although videogame playing may seem to be rather mindless, it is capable of radically altering visual attentional processing."[12]

Subsequently, multiple investigations have confirmed that playing certain videogames confers on the gamer a wide range of diverse benefits, including enhancements in low-level vision, vi-sual attention, and speed of processing, among others.[13] The fact that a number of properly controlled studies have repeatedly dem-onstrated a causal link between videogame playing and the en-hancement of these abilities proves that the videogames, and not any preternatural gifts of the players themselves, are causing this improvement. Nor does the videogame experience have to result only in an immediate advantage in current tasks. A real benefit of playing appears to be the even more impressive ability to improve on how gamers will learn completely new tasks. These newfound talents have subsequent real-world applications. They include, for example, a superior ability to see small details, faster processing of rapidly presented information, higher capacity in short-term memory, increased capacity to process multiple objects simulta-neously, and flexible switching between tasks—all useful skills in a variety of precision-demanding jobs. Laparoscopic surgeons who are habitual gamers turn out to be better surgeons than their non-gaming peers in terms of speed of execution and reliability.[14]

Time spent on videogames is not a simple rehearsal of a specific skill but, remarkably, can be generalized to other situations and a wide range of unforeseen skills and behaviors. It is hardly surpris-ing, therefore, that Nintendo advertises Big Brain Academy as a game that "trains your brain with a course load of mind-bending activities across five categories: think, memorize, analyze, com-pute, and identify."[15] Moreover, one of the promises is that, compared to traditional training methods, the game can be engaging and entertaining.

And it is not just the normal, healthy Digital Native brain that appears to flourish. The evidence is convincing that games can have beneficial, remedial effects over a wide range of impairments, including a reversal of cognitive decline in the elderly. In one study, the researchers trained older adults in a videogame for a total of 23.5 hours.[16] They assessed their subjects with a battery of cognitive tasks, including tests for executive control and visuo-spatial skills, before, during, and after videogame training. The subjects improved significantly within the game but, most important, also showed clear improvement in executive control functions, such as task switching, working memory, short-term visual memory, and reasoning. Specifically, participants trained on the videogame were able to switch between two tasks with less effort or cost to their attention than the control subjects, and showed short-term improvements in recall in the executive function tasks they were tested on before and after the training period.

When used to treat patients with a wide range of brain disorders, it seems, videogames can offer a truly beneficial and enjoyable experience. For example, they have been effective in reducing delusional symptoms in schizophrenic patients after just eight weeks.[17] In a pilot study in adolescents with autistic spectrum disorders, there were visible changes in brain scans in response to emotional words and emotions during a six-week period of prosocial game playing.[18] In the rehabilitation of the victims of motor vehicle accidents with post-traumatic stress disorder, the virtual-reality experience of driving or riding in a car in a computer game improved symptoms and promoted recovery.[19] Videogames catering to specific psychological needs in certain disorders can offer effective complementary treatment options, such as for those with impulse control problems.[20] Meanwhile, neuroscientists have been using popular iPhone games such as Fruit Ninja (where you simply slice fruit in half with your finger) to rehabilitate stroke victims.[21]

Playing videogames could also potentially have positive effects on more abstract aspects of brain function, such as social development and psychological well-being. For example, playing videogames together with parents has been linked with decreased levels of aggression and increased levels of prosocial behavior, albeit only in girls.[22] However, the same research found that the length of time spent gaming, in general, was associated with increased aggression and lower prosocial behavior. Therefore, the beneficial effect here could be due more to the joint activity with parents than to the actual action being played out on the screen. Even gender stereotyping might play a part. The authors speculate that because boys play more videogames than girls, the time the boys spent playing games on their own may have diluted the beneficial effects of time spent playing games with parents. Additionally, they suggest that boys typically play more age-inappropriate videogames than girls, and this may also offset the benefits of gaming with parents.

As we've already seen with social networking, videogame worlds may be a realm where gamers can freely explore their identities.[23] Research shows that tapping into leadership potential in MMORPGs can spill over into workplace potential.[24] Such games could perhaps help develop new organizational training techniques; then again, it could just be the case that a gamer who has the potential to be a leader in a videogame ends up as a leader in the real world, while losers in the real world remain losers in a game. It's still debatable whether videogames serve as a useful lesson for real life or as an escape from it. Games may indeed demonstrate to the gamer that choices are sometimes hard to make, as when gamers are trying to achieve a goal and must weigh consequences, benefits, and the strength of their individual skill set as they decide whether to confront or avoid a problem. On the other hand, experience in the real world will teach that anyway. After all, if there were no difference between real life and gaming, what would be the point of the game in the first place? But if there *is* a difference, would the game experience actually be that useful in terms of real-life applications?

Almost all real-life tasks could be considered dull in comparison to well-designed, highly stimulating games, and this difference may have seriously negative consequences. Kira Bailey and her research group at Iowa State University cautiously note that while some videogames may have positive educational and therapeutic effects, overall their data suggested "that high levels of videogame experience may be associated with a reduction in the efficiency processes supporting proactive cognitive control, that allow one to maintain goal-directed information processing in contexts that do not naturally hold one's attention."[25] Or to put it more simply, gaming could be bad for sustained attention.

While extensive research has shown how action gaming can improve focusing on the screen, this gain may come at a cost. Videogames reward players for quickly modifying their behavior when conflict is experienced, and this specific feature of action games may have differential effects on proactive and reactive control. Think of *reactive control* as the just-in-time type of response to a stimulus that is used only when needed, whereas *proactive control* would be deployed consistently and in anticipation of future stimuli, indicating an individual's capacity to choose what she pays attention to and what she ignores.[26] While high-frequency gamers (playing more than forty hours a week) are well rehearsed in responding instantaneously to suddenly presented stimuli (reactive control), their ability to maintain proactive attention over an entire task is less impressive. Videogames may train an individual to respond rapidly to suddenly presented stimuli, but they may provide no advantage in being able to maintain focus during mundane tasks.[27]

In contrast, other recent work suggests that frequent gamers may be more persistent than infrequent gamers in sticking with complex puzzles involving anagrams and riddles.[28] Frequent videogame players spent more time on unsolved problems compared to infrequent videogame players. These results were taken as proving that videogame use can lead to more perseverance across a variety of tasks. However, once again it could be the case that different character traits are responsible for the crucial difference within an experimental protocol. Gamers may be more competitive than nongamers, and a laboratory assessment task measuring skill of any type will automatically motivate a frequent gamer to want to win. Moreover, the gamers in this study may have seen the puzzle as a game itself and not as a boring task. So the question is still open as to whether frequent gamers will have the ability to pay more attention generally, regardless of the task at hand.

How can we reconcile conflicting conclusions as to whether gaming improves or impairs attention? The answer may lie in the *type* of attention required to be successful at action games. There are a number of taxonomies that attempt to describe the human attention system. *Selective* or *focused attention,* defined as the ability to focus on a specific set of stimuli, is a kind of attention that is typically driven by internal motivations. *Sustained attention,* by contrast, is the ability to maintain vigilance over longer periods of time, and is often required during a tedious activity. While videogames might rehearse and therefore be beneficial to the type of attention requiring the processing of selective stimuli, the maintenance of attention over long periods in the absence of fast-paced moment-to-moment stimulation could well be diminished. So gamers may have a problem not with selective attention but rather with sustained attention.

One interesting question about these impairments in attention is their possible connection to the prevalence of attention deficit hyperactivity disorder.[29] For some, the idea that attentional disorders could be linked to gaming is mere speculation. In a review assessing the impact of digital technologies on human well-being, Paul Howard-Jones concluded, "We do not know [if] the use of digital technology by young children is a causal factor in developing ADHD."[30]

Subsequently, Alison Parkes and her team at the University of

Glasgow surveyed more than eleven thousand children and reported that videogames had no effect on their psychosocial development, including attentional problems.[31] The size of the cohort studied here might seem impressive and hence the findings reassuringly conclusive. But there are some serious drawbacks to the underlying methodology. First, the study investigated children between the ages of five and seven, while almost all the rest of the research literature focuses on older children, who have greater opportunity to play stimulating, violent, or reckless action games not typically available to the very young. Second, the possible symptoms for ADHD were assessed solely by subjective report of the far-from-unbiased parents (hence the unusually large sample size, as data was relatively easy to collect); in contrast, other studies have used more comprehensive, time-consuming, and objective assessment tools. Third, the Glasgow project measured only weekday videogame use, and there may be many more hours of videogames played on the weekend, so the study does not provide a complete picture of total videogame use.

In any event, before we can be sure of a link between attentional problems and gaming, various other issues need to be unpacked. A number of studies have investigated the relationship between excessive Internet use generally and ADHD symptoms.[32] A huge caveat, however, is that gaming and excessive Internet use are two distinct activities: one might be related to ADHD, while the other might not. A further complicating factor is that certain genres of games may have different effects on ADHD. MMORPGs are actually associated with lower levels of impulsivity and ADHD symptoms, but in turn are linked to higher levels of anxiety and social withdrawal.[33] Moreover, the relationship between ADHD and gaming might hinge on the actual frequency of playing, which will not necessarily have been taken into consideration. In addition, any relationship between excessive Internet use and ADHD may be attributable to an addictive state and not to the activity itself. That said, given that so many excessive Internet users are gamers, the relationship between excessive Internet use and ADHD needs to be explored.

Taking all the above considerations into account, and bearing in mind that there is no single "cause" of ADHD, there is still persuasive evidence that excessive gaming can indeed be associated with attentional disorders. In 2006 Jee Hyun Ha and his colleagues investigated large numbers of children in Korea in two stages. The first consisted of screening all participants for Internet addiction disorder and then, from those who screened positive, randomly selecting a smaller group for a thorough psychiatric assessment. Tellingly, the Internet-addicted children used the Internet primarily for Internet gaming. Over half of these youths (who were ages nine to thirteen) qualified for a diagnosis of ADHD.[34] A year later a psychiatric comorbidity survey of more than two thousand Taiwanese high school students aged fifteen to twenty-three reported that 18 percent of students were classified as Internet addicts and that Internet addiction was strongly associated with ADHD symptoms.[35]

As well as the finding that restriction of children's exposure to TV and videogames reduced the likelihood of attention problems in class, a study by Philip Chan and Terry Rabinowitz at Rhode Island Hospital found that if teenagers played videogames for more than one hour per day, they displayed more features of ADHD, including inattention.[36] However, the authors highlighted the now familiar chicken-and-egg problem: "It is unclear whether playing videogames for more than one hour leads to an increase in ADHD symptoms, or whether adolescents with ADHD symptoms spend more time on videogames."[37]

While there is a significant association between the level of ADHD symptoms and the severity of Internet addiction in children, it also appears that the presence of ADHD in a child might predict the likelihood of developing gaming addiction. In a study of young people with and without ADHD, ages six to sixteen years, there were no differences in the frequency or duration of gaming between the two groups.[38] However, the ADHD group had significantly higher gaming addiction scores, indicating that ADHD children may experience gaming activity with more intensity than non-ADHD children and thus may be particularly vulnerable to gaming addiction. So if Internet use and obsessive gaming are influencing each other, it may not be that one is causing the other, but rather that both are symptomatic of the same single common brain state: two sides of the same mental coin. A clue as to what that brain state might be comes from looking a bit more closely at a medication used to treat ADHD.

Methylphenidate, perhaps best known by one of its brand names, Ritalin, is a stimulant drug given widely to treat attentional disorders. In the United Kingdom, the number of prescriptions for methylphenidate soared from 158,000 in 1999 to 661,463 in 2010.[39] In the United States, Benedetto Vitiello of the National Institute of Mental Health documented stimulant prescriptions between 1996 and 2008 and found that the number of prescriptions for children younger than nineteen increased significantly during that twelve-year period.[40] Those aged six to twelve had the most prescriptions, but teens aged thirteen to eighteen had the biggest *increase* in prescriptions. A similar trend was found in Australia, where the use of stimulant drugs to treat ADHD in children has escalated dramatically, with prescriptions for Ritalin and its equivalents up 300 percent between 2002 and 2009.[41]

Of course, it could be that these colossal increases in prescriptions across three different continents have nothing to do with an increase in ADHD itself but owe more to a current clinical trend to medicalize a particular behavior and/or a greater willingness to prescribe medication for the condition.[42] Nonetheless, the current association between ADHD medication and abnormally short attention spans brings into play our old friend dopamine, as methylphenidate results in an increase in this chemical messenger in the brain. It has proved a continuing riddle to neuroscientists why such a drug should be effective in treating a short attention span.

When dopamine goes to work in the brain, you become more aroused, more excited. The apparent paradox of a stimulant drug such as Ritalin effectively combating hyperarousal can be explained by its ability to desensitize dopamine's normal chemical targets. As we've discussed already, the interaction of these chemical targets (receptors) with their respective brain neurotransmitter resembles a molecular handshake. But if the handshake is persistent and strong, then the hand (the receptor) will become numb, less sensitive (desensitized). The result is that the dopamine in the brain will be less effective and you will be less hyperactive. In an individual who does not have ADHD, the drug can prolong attention span, which could be viewed as a desirable "cognitive enhancement."

Modafinil, a novel wakefulness-promoting agent, has a pharmacological profile similar to that of conventional stimulants such as methylphenidate. Psychologist Trevor Robbins and his team in Cambridge were interested in assessing whether modafinil might offer similar potential as a cognitive enhancer in those who were perfectly normal.[43] Sixty healthy young adult male volunteers received a single oral dose of either a placebo (an inert substance that they thought could have beneficial effects) or of modafinil prior to performing a variety of tasks designed to test memory and attention. Only modafinil significantly enhanced performance on various cognitive tests, including visual pattern recognition memory, spatial planning, and reaction time. The subjects also said that they felt more alert, attentive, and energetic on the drug. A further effect seemed to be to reduce impulsivity. So might drugs such as modafinil give us an insight into the link between ADHD and excessive gaming?

In 2009 associate professor of psychiatry Doug Hyun Han and his team at the University of Utah prospectively studied a large number of teenagers, the great majority of whom were male. The subjects all had a history of ADHD, as well as track records of excessive use of videogames. The idea was to examine whether both videogame play and methylphenidate increased dopamine release in a way that could enable the teenagers to concentrate better. Han administered Concerta XL (similar to Ritalin) and followed up the performance of the subjects after eight weeks. There was a reduction in Internet addiction scores and total time of Internet use, indicating that methylphenidate could reduce such obsessive behavior in subjects with co-occurring ADHD and excessive gaming. Although the authors did not clarify how much of the Internet activity was gaming, they came to the fascinating conclusion that if ADHD and gaming really are two sides of the same coin, same brain state, then "Internet videogame playing might be a means of self-medication for children with ADHD."[44]

If videogames are a kind of self-medication for those suffering from ADHD, the most obvious common factor is excessive dopamine release in the brain, in turn related to addiction, reward, and arousal. Paul Howard-Jones at Bristol University has even suggested that this process could be harnessed by allowing children to play videogames; they would thereby become more aroused and be cognitively enhanced in the classroom.[45] So perhaps, under the right conditions, videogames might prove a valuable tool for teachers. Yet while the amounts of endogenous dopamine released naturally within the brain as a consequence of gaming would probably not lead to the same level of receptor desensitization that could occur with usual doses of modafinil or Ritalin, do we really want students to be in a permanent state of high arousal? It would surely not be that different from giving them low doses of amphetamine.

Most immediately, it seems there is a clear link between gaming and attention generally. Although *selective* visual attention for focusing on a screen object or avatar might be improved in the short term with gaming, it could be to the detriment of the all-important *sustained* type of attention over the longer term, the kind of attention needed to reflect and to understand something in depth.

Moreover, the implication of dopamine as a central player in the brain of the gamer might be providing a truly helpful insight into understanding the appeal of the activity, compared to real life. But could a mindset used to experiencing reliable if not easy rewards, also be one inclined to aggression and recklessness?

VIDEOGAMES, AGGRESSION, AND RECKLESSNESS

I t seems incredible that the prototype videogame Pong first appeared as long ago as 1975. But it wasn't until the 1990s that games such as Double Dragon and Mortal Kombat introduced more violent acts into play. The pictorial resolution of these early games was measured in polygons per second, and can be a good indicator of how fast this technology has developed. For example, the resolution of the first PlayStation model was 3,500 polygons per second, but by 2001 the original Xbox released had a graphics quality of 125 million polygons per second. Now current electronic games have astonishing graphic resolutions in excess of 1 billion polygons per second! As a result, the screen portrayal of violence in videogames has become more detailed and vivid. Players are now exposed to multiple ways of killing, and witnessing death in cyberspace more frequently and much more vividly than ever before.

The issue of just how graphic videogames might have nasty consequences revisits the now familiar argument that cyber-based activities in general, and gaming in particular, are being disproportionately demonized, while older technologies, such as TV, have always been just as detrimental. Not so. Hanneke Polman and her team at Utrecht University explored the difference between playing a videogame and a more TV-like experience of passively watching violent videogames.[1] After being exposed to the videogames, the students had two free-play sessions, after which they completed a questionnaire on aggressive behavior. Acts were labeled aggressive only if the intention was considered hostile. The Dutch team found that, particularly for boys, actively playing a violent videogame led to more aggression than just passively watching the same violent videogame.

The crucial difference between passively observing media violence and playing a violent videogame is, most obviously, the interactivity. In many games, the player is "embedded" in the game and uses a handheld controller that enhances the experience and thus could escalate aggressive feelings. But then again, violent videogames could affect behavior in the real world only if the player ended up confusing the two. If someone only played Super Mario Bros., would we be concerned they would start to believe in turtle shells that can knock people out and feathers that make you fly?

This is an *ad absurdum* argument. First, no one is claiming that violent videogames are the sole and exclusive influence on any individual's actions. Humans don't exist in a vacuum. Even the most avid gamers live a life beyond their consoles: they go to school and they learn from their parents and peers. Second, comparing cartoon violence to graphic, hyperreal violence is a stretch. People are less likely to be influenced by a game completely devoid of reality such as Super Mario Bros. compared to one that mimics reality, such as Grand Theft Auto V. Videogame violence taps into established mental schemas we already have around aggression and real-world violence. Turtle shells, feathers, and being able to fly don't have those established toeholds in our minds, whereas strangers being potential aggressors, and our subsequent feelings of hostility and distrust toward them, do. Additionally, researchers looking into this subject have questioned the true level of violence in cartoon-style games that are geared toward children, such as Super Mario Bros.[2] The majority of studies have focused on highly graphic and realistic human-on-human character violence, with modern games featuring vividly detailed and gruesome acts such as decapitation—and a really important point is that this type of realistic violence *does* appear to impact on levels of subsequent aggression.

Elly Konijn and her group at the University of Amsterdam tested the hypothesis that violent videogames are especially likely to increase aggression when players identify with violent game characters.[3] A large group of adolescent boys were randomly assigned to play a videogame with realistic violence, a fantasy videogame, or a nonviolent videogame. Next, they competed with an ostensible partner on a reaction time task in which the winner could blast the loser with loud noise through headphones, which served as the measure of aggression. Participants were told, wrongly, that high noise levels could cause permanent hearing damage. As expected, the most aggressive participants turned out to be those who played a violent game and wished they were like a violent character in the game. These participants used noise levels that they believed were loud enough to cause permanent hearing damage to their partners, even though their partners had not provoked them. The results suggest that identifying with violent videogame characters makes players more proactively aggressive, even after controlling for habitual videogame exposure, trait aggressiveness, and sensation seeking. Players were especially likely to identify with violent characters in realistic games and in games in which they felt immersed. So it would seem the boys were not simply rehearsing stereotypical violent responses but were taking on a more generally adversarial mindset.

Then again, there are those who still question whether videogames could ever actually lead to violence. They argue that the gaming experience cannot actually be harmful because humans have an inherent ability to recognize right and wrong. But we've seen time and again in these pages how we are shaped by our individual experiences and how the human brain always adapts to its environment. If that environment for many hours of the day is one of intergalactic warfare or of supernatural heroes with magic powers, then that fiction might increasingly inform the brain's understanding of reality, and ultimately of good and bad. And in fact this seems to be the case.

Recent evidence suggests that despite a gamer's awareness that the game world is not real, he or she still has a real human response to game events. Andrew Weaver and Nicki Lewis at Indiana University designed a project to discover how players make moral choices in videogames, and what effects these choices have on emotional responses while playing.[4] Seventy-five participants filled out a "moral foundations questionnaire" and then played through the first full act of the action videogame Fallout 3. The majority of players arrived at moral decisions and behaved toward the nonplayer game characters they encountered as if these were actual interpersonal interactions. The gamers felt guilt when they engaged in an immoral act toward a (nonhuman) videogame character in the game, but, tellingly, this guilt didn't affect their level of enjoyment. It is surely strange that people feel guilt toward a character they know is not a human and doesn't really exist. Moreover, even if for the time being the decisions were "moral," the enjoyment alongside the culpability suggests that, while feeling guilty may well imply a certain level of empathy, ultimately there is still an interesting decoupling between understanding someone's suffering and caring about it sufficiently to modify your actions.

Yet, the same argument could be made about books, you might say. We can feel an emotional attachment to, and empathy with, characters, but this in no way lessens our enjoyment of the novel itself. How are videogames any different? Well, beyond the opportunity for escapism in both cases, the enjoyment of books could be due to the insight the reader gains from experiencing the lives of others at different times and places, giving them the opportunity to adjust their views and perhaps serving as a trigger for new ideas. No such claim has been made for videogames, where, as we saw previously, much of the pleasure comes from the release of dopamine in a directly interactive and fast-paced experience that does not occur when reading a book. Most significant is that, however gripping a novel is, no one would conflate it with the real world around him or her, as might be possible with videogames. Through your avatar, you can live another life. Despite knowing that this world is a fiction, gamers *do* appear to conflate fantasy with reality in violent videogames.[5]

Craig Anderson, professor and chair of psychology at Iowa State University and a leading researcher in the field of videogame violence, is concerned that while violent games do not cause extreme, criminal-level violent behavior, they do enhance low-level aggression. He is convinced that he and others working in this area

now have a clear picture of how media violence increases aggression in short- and long-term contexts. Immediately after exposure to media violence, there is an increase in aggressive behavior tendencies because of several factors. 1. Aggressive thoughts increase, which in turn increase the likelihood that a mild or ambiguous provocation will be interpreted in a hostile fashion. 2. Aggressive affect increases. 3. General arousal (e.g., heart rate) increases, which tends to increase the dominant behavioral tendency. 4. Direct imitation of recently observed aggressive behaviors sometimes occurs.[6]

Anderson's suggestion is that the link between aggression and gaming is an indirect and generalized association. Indeed, it's quite plausible that subconscious leanings toward violence could be transformed into overtly conscious ones via gaming and through repetition could become automatic, the default mode. It is the rehearsal, the repetition, that is all-important, as the player is immersed in the fantastical narrative played out over and over. Compared merely to observing a violent scene, in the course of playing an actual game you have a persona whose aggressive actions are rewarded by the game, triggering a dopamine rush in your own brain; therefore your aggressive mindset becomes the norm. The individual who has engaged in violent gaming could lose self-awareness and insight because the tendency to an aggressive disposition has become a strong habit.

We have already seen how visionary psychologist Donald Hebb stated more than seven decades ago that neurons that "fire together wire together." More recently, videogame researcher Douglas Gentile has echoed this theme, pointing out that "whatever we practice repeatedly affects the brain and if we practice aggressive ways of thinking, feeling and reacting, then we will get better at those."[7] The violent content of computer games could desensitize players to violent behavior toward others, in part by lowering the threshold of response to provocation and through the dwindling of empathy with other people. For example, if someone bumps into you in the corridor, you could overreact with a hostile "Who do you think you're shoving!"

In a recent study, Youssef Hasan and his group at the University Pierre-Mendès-France showed that violent gaming does indeed increase expectations that another will act with hostility or aggression, probably as a result of repetitive experience in a game with hostile characters.[8] French college students played either a violent game or a nonviolent game for just twenty minutes. Afterward, they read ambiguous story plots about potential interpersonal conflicts, and listed what they thought the main characters would do, say, think, or feel as the story continued. Aggression was measured using a competitive computer game in which the winner could apparently blast the loser with loud noise through headphones. Results showed that the violent videogame players expected more aggressive responses from the main characters presented in the story. Moreover, they chose significantly louder and longer blasts of noise for their human opponents in the game. As predicted, videogame violence increased the hostile expectation bias, which in turn increased actual aggression. What will be the longer-term implications of this state of affairs?

One suggestion is that there could actually be some positives. For example, violent videogames may provide a safe outlet for aggression and frustration.[9] In this spirit, research currently being led by Cheryl Olson and her team at Massachusetts General Hospital's Center for Mental Health and Media indicates that violent games help students deal with stress and aggression. Apparently, more than 45 percent of boys and 29 percent of girls use violent games such as Grand Theft Auto IV as a safety valve for their anger.[10] But there is little evidence that violence is an internally generated biological imperative akin to hunger or sleep—a drive that builds up in the body come what may, as a natural need that sooner or later must be met. Furthermore, anger is not the same as aggression, although the former might sometimes lead to the latter. In any case, it could be that there are more effective ways to help someone cope with anger than providing an opportunity for violence, however simulated.

The only "proof" that violent games might have positive effects, according to Olson and many other gaming aficionados, seems to be that the violent crime rate is going down while the popularity of violent videogames has increased. But decreases in the crime rate are most likely to be the product of a host of complex socioeconomic factors. Most important, no one has ever actually demonstrated a direct link between violent videogames and a decrease in actual violent crime, or indeed suggested the converse, that such games directly drive players to go out on a rampage.

However, the change toward a more aggressive disposition as a result of videogames does seem to be a definite global phenomenon across different cultures. A recent longitudinal study designed to explore the long-term effects of violent games on the mentality of American and Japanese school-age young people has shown that in as little as three months, high exposure to violent videogames increases physical aggression, such as punching or kicking someone or getting into physical fights.[11] Other recent, similar studies in Germany[12] and Finland[13] have demonstrated similar effects.

Although the systematic study of videogames is relatively new, the evidence seems strong for a link between playing videogames and an aggressive mindset. The most comprehensive meta-analysis to date has drawn on 136 papers detailing 381 independent tests of association conducted on a total of 130,296 research participants, finding that violent game play led to significant increases in desensitization, physiological arousal, aggressive cognition, and aggressive behavior, while prosocial behavior decreased.[14]

As is the way in the peer-reviewed scientific literature, this report was immediately criticized for a number of methodological flaws, in particular a bias in the selection of studies included, as well as allegedly trivial size effects.[15] The original authors, Brad Bushman and his colleagues, were able to refute this accusation and denied that there was any evidence of bias in their selection of data[16]. They also countered that the effects observed, far from being trivial in size, were bigger than many effects deemed sufficiently large to warrant action in medical domains. So the main argument against the potentially detrimental effects of violent gaming comes down to one of detail (the real-world implications of those effects, their magnitude, and the methodology for evaluating them), but *not* to whether any exist in the first place.[17]

Beyond aggressive behavior toward others, violent videogames clearly *do* have a demonstrable effect on the brain and body. Research has linked violent videogames to changes in the fight-or-flight system that has evolved to prepare the body for action by pumping blood around the body more quickly, putting digestion on hold, cooling down the skin with sweat, and so on. It seems that players can become habituated to this adrenal rush, such that living through a realistic violent experience will no longer trigger as strong a reaction.[18]

Nicholas Carnagey, a psychologist at Iowa State University, demonstrated that brief exposure to violent videogames influences activation of the part of the nervous system that usually gets your heart racing automatically.[19] The subjects played a violent or a nonviolent videogame for twenty minutes and immediately after playing the game viewed a ten-minute video clip of actual real-world violence (not Hollywood reproductions)—for example, a prison fight in which a prisoner was stabbed repeatedly—while heart rate and skin conductance were measured. Those who had played the violent videogame demonstrated less change in heart rate and less sweating of the palms while watching the video, compared to those who played the nonviolent videogame. The violent videogame had made the subjects less affected and upset by the real-world aggression.

The consequences of such physiological desensitization could be quite significant: when individuals are desensitized by violent videogames, they are less likely to aid a victim of violence.[20] In one particular study by Brad Bushman and Craig Anderson at Iowa State University, subjects played one of the videogames before a fake fight was staged outside a laboratory toward the end of the study. Compared to those participants who had played the nonviolent videogame, participants who had played the violent videogame were less likely to report hearing the fight, judged the event as less serious, and were slower to respond when they did offer help.

Perhaps not surprisingly, playing violent videogames has corresponding effects that can be observed in the brain itself. Brain activity monitored during game play shows that there are definite neuronal correlates to real-life behavior. Investigators recorded the brain activity of experienced gamers, who normally played an average of fourteen hours per week, while they played a first-person shooter game.[21] Watching violent scenes caused a change in activity in certain areas of their brains, and specifically one particular area, the rostral anterior cingulate. This area is normally active during detection of discrepancies in incoming information, such as in the Stroop test, when reaction time is slower because the name of a color (e.g., blue) is printed in a color not denoted by the name, such as red. Gaming was also correlated with the deactivation of the amygdala, a brain region normally linked to emotionally charged memory, such that decreased activity in this area would lead to the suppression of fear and an overall drop in emotion. The brains of the gamers were therefore less sensitive and less emotionally reactive to discrepant actions, such as sudden violence. It's important to note that the activation pattern reflected a sequence of the individual's own brain-environment interaction rather than just merely registering what was going on.

In a second imaging experiment, regular male gamers played a first-person shooter game and their actions in the game and their corresponding brain scans were analyzed.[22] Results showed that areas of the brain linked with emotion and empathy (again the cingulate cortex and the amygdala) were less active during violent gaming. The authors suggest that these areas must be suppressed during violent gaming, just as they would be in real life, in order to act violently without hesitation. Furthermore, there was activation of areas associated with aggression and cognition, paralleling the activation that occurs during real-life violence.

Does this mean that the brain can't tell the difference between a virtual act of violence and a real-world act of violence? This is the same as asking whether individuals (who, after all, *are* their brain) can make such a distinction. We've already seen that gamers can conflate reality and the virtual world. If the opposite were true, if there was some kind of neuronal reality check, it's hard to see where and how it would operate in the physical brain as a mechanism capable of bestowing objectivity independent of all other brain processes. If, as I've been suggesting, the mind is the personalization of the brain through personalized neuronal connectivity, each of us will, in any case, have unique and very different views of an external reality. It would be foolhardy to assume that the human brain *always* knows the difference between fantasy and reality. Neuroscientist Rodolfo Llinás of NYU Medical Center has gone so far as to argue that our default consciousness is internally generated, modified only to a greater or lesser extent by an intermittent input from an external reality.[23] Meanwhile, the extreme idea that all reality is illusory and external objects exist only when they are perceived goes back centuries to the philosopher George Berkeley. Here is not the place to discuss the nature of physical reality, but suffice it to say that there is no automatic switch in the brain for detecting it, nor for assuming that it is a simple concept in the first place, easily distinct from the imagination, that we can take for granted, let alone define.

Although we have focused here on heavy gamers, those who might be obsessive if not actually addicted, the picture that is emerging is one of a clear relationship between violent gaming and increases in aggressive thoughts, feelings, and behavior. But what does this actually mean for life beyond the screen? We know from multiple well-designed laboratory studies that playing violent videogames can make us respond more aggressively. But how long these effects last and whether they translate inevitably into real-world situations remain unclear.

Our exploration of videogames started with the idea that, in playing games, we rehearse many of the skills useful for survival in the real world. The possible link between aggression and gaming is still debated even after twenty-five years of research, because terms such as "aggression," "aggressive behavior," "anger," "hostility," and even "aggressive cognition" are often poorly defined, measured indiscriminately, and used interchangeably. But above all we need to distinguish "anger," "aggression," and "violence." There is no evidence that videogames lead directly to criminal-level violence, but a large body of data strongly indicates that they do induce an aggressive disposition in everyday life. This is particularly worrying in light of recent statistics that violent videogames account for approximately 60 percent of videogame sales.[24] Moreover, at the time of writing, the top five most popular videogames (Grand Theft Auto V, Batman: Arkham Origins, Assassin's Creed IV: Black Flag, Call of Duty: Ghosts, Battlefield 4) are all extremely violent in content.

As we've seen throughout these pages, humans are mandated by evolution to adapt to the environment. Children have always learned best by observing behavior and then trying it out for themselves. The consequences of these experimental forays influence whether they repeat the behavior or never do it again. All violent media, regardless of type, have the potential to teach specific violent behaviors, as well as to color the circumstances when such behaviors seem appropriate and useful. In this way, violent behavioral scripts are learned and stored in memory. Videogames provide an ideal environment in which to learn violence because they place players in the role of the aggressor and often reward them for successful violent behavior. Games allow players to rehearse an entire narrative, from provocation to choosing to respond violently to resolution of the conflict. Players are incentivized to reenact these scenarios repeatedly and for long periods of time in order to improve their scores and advance to higher levels. Inevitably, this repetition increases their effectiveness and the likelihood of such behavior being repeated. In turn, aggressive behaviors will be adopted. The potential shift to a more aggressive behavioral pattern and attitude over time could affect society and what we expect from one another, possibly lowering our expectations of respect and tolerance and increasing our distrust of others and our perceived need for self-preservation.

Any surge in hostility implies a decrease in normal self-control and an increase in recklessness heedless of the consequences. If a neuroscientist is asked to say something about excessive risk taking, she may well start by pointing to neurological syndromes where brain malfunction is characterized by taking too many risks—think back to Chapter 8 and the case of Phineas Gage, he who was exceedingly capricious, childish, and impatient at being restrained. Recall too that this is a behavioral profile seen in other populations, such as obese people, gamblers, schizophrenics, and of course children. These groups are very different, but they all share a preference for the here and now that trumps the consideration of long-term consequences. Anyone who overeats knows what will happen, but for those with a high body mass index (weight relative to height) the thrill of the taste of the food trumps the consequences that it will pile on the calories. Similarly, research has shown that obese people are more reckless in gambling tasks, and are comparable to compulsive gamblers for whom the thrill of the horse passing the finishing post, or the roll of the dice, trumps the consequences that they may well lose all their money.[25] But then, what of schizophrenics who may be neither obese nor compulsive gamblers?

A detailed excursion into schizophrenia is outside the scope of our current discussion, but the main feature to flag is that schizophrenics place a higher emphasis on the outside sensory world, which they often think is imploding in on them. They think that outsiders can see and hear their thoughts, since there is no firewall between their brains, or rather their minds, and the incoming flood of sensory stimulation impinging on them. We've seen that, as we develop, the sensory world is overtaken by a more cognitive one, where personalized associations, *meaning,* dominate our interpretations of the world. In schizophrenia, this transition is far less emphatic and, as the senses remain overly dominant, those with schizophrenia are more easily distracted by novel stimuli and have shorter attention spans.[26] Those with schizophrenia also struggle with proverbs and metaphorical thinking, which we saw previously with the characteristically literal interpretation of the statement "People who live in glass houses mustn't throw stones" as signifying "If you live in a glass house and I throw a stone at it, then your house will break." Schizophrenics have trouble understanding one thing in terms of something else because the ability to make such associations is usually based on a robust functional connectivity between networks of neurons, a connectivity that grows and is personalized throughout life.[27]

Another group of people who see the world literally and take it at a sensory face value are children. A young boy or girl instructed not to cry over spilt milk might look around in surprise at the absence of an overturned glass. Young children can be compared to adults with schizophrenia in that they have shorter attention spans, are more readily distracted, and significantly, are also more reckless. They too have an underactive prefrontal cortex, which fully matures only in the late teenage years or even early twenties.[28]

As we saw earlier, the common factor underlying obesity, schizophrenia, gambling recklessness, and childhood is how the sensory present trumps the long-term consequences: the press of the here and now environment is unusually paramount. And this suppression of the past and future in favor of the present moment seems to be related to an underactive prefrontal cortex. Does this mean that, despite all the health warnings in the earlier chapters against regarding specific brain regions as independent mini-brains, the prefrontal cortex is indeed some kind of HQ for cognition and loftier thoughts beyond the moment? Not at all. Far from being a kind of autonomous super mini-brain, the prefrontal cortex has more inputs to all other cortical areas than any other region of cortex, and therefore plays a key role in operational brain cohesion. So, if this pivotal area is underactive for whatever reason, there could be a profound effect on holistic brain operations, which are normally functional for accessing memories and planning ahead. One interesting effect of damage to the prefrontal cortex can be "source amnesia," where memory is still intact but is more generic and is removed from a specific context or episode.[29] The patient is not linked to a continuous narrative of particular events, but is more in an ill-defined, hazy present.

When dopamine accesses the prefrontal cortex, it inhibits the activity of the neurons there,[30] and so recapitulates in some ways the immature brain state of the child, or indeed of the reckless gambler, the distracted schizophrenic, or the food junkie. Just as children are highly emotional and excitable, adults in this condition are also more reactive to sensations rather than calmly proactive. Small wonder that this much-cited transmitter heightens arousal and arousal is often linked to pleasure whether it be in extreme sports, drugs, sex, or rock and roll. It's a brain state dominated by the heightened sensational moment for the passive recipient. So, when you "blow your mind," you temporarily suspend access to the personalization of neuronal connections developed over an individual lifetime that characterizes your proactive uniqueness. Now, for the time being at least, those connections are not being accessed fully, thanks either to psychoactive drugs or to an environment that has little cognitive content because the senses are being rapidly and powerfully stimulated, as in the context of sports or sex or raves.

How might this scenario apply to videogames? A character you've just shot in a videogame can become obligingly undead the next time around. Perhaps the biggest difference between videogames and real life is that in games, actions do not have irreversible consequences. You can afford to be reckless in a way that would have dire results in the three-dimensional world. The consequence-free nature of gaming is a basic part of its ethos (remember that one of Nicole Lazzaro's essential criteria for a successful game, is to "suspend consequences").[31] Depending on the game, sometimes you'll even be rewarded for behaving recklessly while playing. This parallel world not only facilitates recklessness but, depending on the game, sometimes even rewards it. This cyber-based irresponsibility can have serious effects in the real world. After playing a videogame where reckless driving—crashing into other cars, driving on the sidewalk, driving at high speed—was part of the game, gamers were more likely to behave recklessly and take risks in a simulated driving situation.[32] One longitudinal study found that playing violent, risk-encouraging videogames, including the driving game Grand Theft Auto, was associated with self-reports of risky driving, even after controlling for other variables that influence this type of behavior.[33] Specifically, gaming was associated with vehicle accidents, being stopped by the police, and unsafe driving habits, including speeding, tailgating, and the willingness to drink and drive.

With modern videogames, the mere experience of recklessness itself can be fun. We've already seen that games often provide a fast-paced, exciting experience, one that is associated with high levels of dopamine in the brain.[34] Dopamine is well known to inhibit the prefrontal cortex. So would the brains of gamers display less activity in this crucial brain region? Several studies have indeed linked heavy gaming to decreases in prefrontal cortex activity.[35] A recent report from China found structural abnormalities in the prefrontal cortex in the brains of Internet addicts (and, as we saw in Chapter 14, the majority of Internet addiction studies involve individuals whose main addictive behavior is gaming), which suggests that Internet addiction might result in brain structural alterations.[36] The study involved scanning the brains of adolescents who played on average ten hours of online videogames per day for nearly three years and comparing the results with scans of comparable subjects who played fewer videogames. In the heavy gamers, the scans showed abnormalities in the brain's white matter, the fibers connecting brain regions involved in emotional processing, attention, decision making, and cognitive control.[37]

Similar microstructural abnormalities have been observed in the brains of people addicted to substances such as alcohol and cocaine. In addition to reduced activation in the prefrontal cortex, recent research into videogame addicts has shown an alarming reduction of activity in regions of the brain associated with visual and auditory processes.[38] The researchers suggest that extended gaming can diminish the responsiveness of the visual and auditory regions of the brain. Perhaps too much gaming in a visually and auditorily stimulating world reduces our reaction to the relatively dull real world because our brains have recalibrated to the videogame world that now seems the norm.

Consider the following possible cycle of events involving someone who plays action videogames. The experience of a fast-paced,
vivid, interactive screen experience is arousing, hence dopamine is
released. The dopamine inhibits the prefrontal cortex, thereby putting the brain into a mindset where the here and now trumps consideration of future consequences, making the fast-paced sensations
of the screen even more appealing in comparison with the slow,
unexciting real world. As the gamer continues, more dopamine is
released, desensitizing its receptors. Now more dopamine is needed
to create the same level of arousal as initially experienced, so the
behavior that produced the dopamine surge is perpetuated to a
greater or lesser extent. In some 10 percent or so of individuals this
cycle will be extreme enough to be regarded as addictive or obsessive behavior.

We may now be living in an unprecedented era where an increasing number of people are rehearsing and learning a new default
mindset for negotiating the world: one of low-grade aggression,

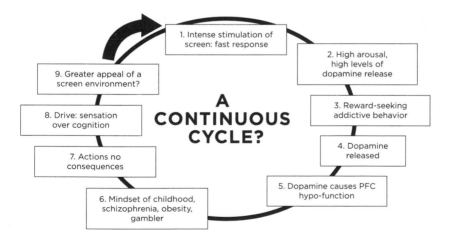

Figure 15.1   A continuous cycle of stimulation, arousal, and reward in addiction that
could account for a compulsion to play games. Typical gaming responses are fast and
exciting, leading to a higher level of arousal and release of dopamine. Dopamine also
underlies rewarding experiences and addiction, so the behavior continues and yet
more dopamine is released. This excessive dopamine will inhibit the prefrontal cortex,
leading to an emphasis on the here and now and a disregard for future consequences.
Playing the game meets this drive for more here-and-now sensory experience particularly well, and so the cycle continues.

short attention span, and a reckless obsession with the here and now. But although excessive gaming may ramp up arousal levels and feelings of reward, it does so in the cognitive context of the Internet game. And this simulated context can perhaps become the new narrative that, in extreme cases, substitutes for the less simple, less successful, less fun story line that is the player's real life.

THE *SOMETHING* ABOUT
SURFING

"**I** wanted something that expressed the fun I had using the Internet, as well as hit on the skill, and yes, endurance necessary to use it well. I also needed something that would evoke a sense of randomness, chaos, and even danger. I wanted something fishy, net-like, nautical."[1] These reminiscences are from librarian Jean Polly, who claims to have been the first to have used the term "surfing" in 1992, as she "cast about for a metaphor" for an article title. But many find this version of events hard to believe. The term is more likely to have evolved from TV channel surfing, as an ironic commentary on how unsporty, safe, and inactive flicking a remote at a television is, compared to the actual riding of the real waves. Alternatively, perhaps channel and Internet surfing resemble actual surfing in that neither of the electronic surfers has much interest at all in what is going on at the deeper levels but just enjoys going along for the ride, wherever it takes them. In any event, the very word "surfing" conjures up excitement, health, youth, and speed as you skim effortlessly across sites, film clips, and facts. It is an activity that is unique to cyberculture.

For the first time ever a vast mass of humanity has easy access to an effectively infinite amount of information via search engines and websites: we can see any backyard in the world via sites such as Google Earth and, if need be, get instantaneous updates on world events as they unfold. Traditional notions of space and time no longer have the same relevance and no longer impose the same constraints on our lives, while most governments trying to monitor their state media no longer have unfettered control over what their citizens can access. Then there is the darker side of surfing: the far less savory opportunities, for example, to learn how to make an improvised explosive device, determine the most effective way to commit suicide, or, unbelievably, find the best method for cooking human flesh. Anyone anywhere can access such sites.

This free-of-charge, casual, rapid acquisition of information even applies to more formal education, with lessons and lectures from all over the world. Since 2001, the Massachusetts Institute of Technology, for example, has made openly available on the Internet the materials for nearly all of its courses, while more recently the Khan Academy has set up twenty-seven hundred high-quality microtutorials on the Web (www.khanacademy.org), and computer games developed by Marcus du Sautoy, a mathematician at Oxford University, are enabling children to engage with complex problems that people would have once said were far too advanced for them.[2]

But surfing can involve much more than formal learning. "Without Google and Wikipedia I'm stupid, not just ignorant."[3] So claimed journalist and visiting Harvard researcher John Bohannon, who went on to speak of the "Google effect," the phenomenon where the Internet becomes a personal memory bank, replacing the collective efforts of family members as a primary source of recall. Bohannon even goes so far as to suggest that many had "made the Internet their husband and wife," a vivid turn of phrase describing how some people now assume Google will complement their memory processes in a way that previously a spouse might have done. Is Bohannon just a weirdo speaking in hyperbole, or has he put his finger on a growing trend?

Bohannon's concerns about the Google effect were based on the results of experiments devised by Betsy Sparrow and her collaborators, Daniel Wegner of Harvard and Jenny Liu of the University of Wisconsin. Their findings illustrating this phenomenon and its impact on cognitive performance made the headlines in 2012 with a paper in the high-impact journal *Science.*[4] Participants read simple statements such as "An ostrich's eye is bigger than its brain." One group of subjects was then tested for their recall of the statements when they believed these had been saved (i.e., the statements would be accessible to them later, as in the case of the Internet), the other when they believed the statements had been erased. Perhaps not surprisingly, subjects didn't learn the facts so well when they believed the information would be readily accessible later. They performed worse on the memory test than the group who believed the information was no longer available and therefore had had to rely on their own cerebral resources from the get-go.

Before we go any further and talk about the impact of Google on memory, we need to sort out the different types of memory that may or may not be affected.[5] *Nondeclarative* memory (or *implicit* or *procedural* memory; the terms are interchangeable) involves the remembered skill set that enables you to ride a bicycle or learn to swim; this type of recall wouldn't be affected by a reliance on Google for summoning up facts. The other kind of memory is known as *declarative* or *explicit* memory, where the process of active recall is either episodic or semantic. *Episodic* memories have specific time-space coordinates and hence can be linked to many other different events and facts that are personal to each individual episode. So, for example, although the September 11 attack on New York's World Trade Center took place at a specific time and location, the actual memory of it will be very different for each of us, depending on our own circumstances and personal history as well as the individual contextual framework in which it was embedded.

In contrast, Sparrow's experiments were dealing with mainly *semantic* memory: objective, stand-alone facts of the type that many would now argue no longer have to clog up our synapses because they can be accessed externally. Although only you can access your personal memories, the idea is that Google, or any other search engine, could eventually act as an outsource for this type of recall of objective facts.

Sparrow devised a subsequent test to explore whether there might be a difference between memory for the information itself and memory for where the information can be found. When asked to remember folder names, subjects did so with greater success rates than when asked to recall the trivial factual content itself. Analysis revealed that people do not necessarily remember *where* to find certain information when they can remember *what* it was; conversely, they tend to remember where to find information when they can't remember the information itself. Sparrow and her colleagues summed this up:

> The advent of the Internet, with sophisticated algorithmic search engines, has made accessing information as easy as lifting a finger. No longer do we have to make costly efforts to find the things we want. We can "Google" the old classmate, find articles online, or look up the actor who was on the tip of our tongue.[6]

This new strategy will swiftly leave its mark on the brain. Gary Small and his colleagues at UCLA studied twenty-four middle-aged individuals, of whom twelve had minimal Internet search engine experience (the Net-naive group) and twelve had more extensive experience (the Net-savvy group).[7] The scientists scanned the brains of these subjects during a novel Internet search task and during a control task of reading text on a computer screen that had been formatted to simulate the prototypic layout of a printed book.

While the brains of the two groups showed similar patterns of activation during the text reading task, the activation patterns were markedly different during the Internet search task. The brain scans of the Net-naive group showed an activation pattern similar to that of their text reading task, whereas the Net-savvy group demonstrated significant increases in activity in additional regions that control decision making, complex reasoning, and vision. Yet, amazingly, after only five days of spending a few hours on the Internet, the erstwhile naive group was showing brain activation patterns similar to those of their savvy counterparts. Once again, we can see the powerful adaptability of the human brain. However, it's not clear whether this seemingly efficient change to the new environment of the Internet is such a good thing. The new brain patterns indicated a switch in strategy from actually reading what was displayed to fast searching, in turn suggesting that success in a Google search depends not on detailed scrutiny or in-depth reflection but instead on fast evaluations at face value.

Of course, using dictionaries, log tables, and encyclopedias also requires quick acts of evaluation. Unlike the Google effect, however, these more traditional resources have never posed a comparable threat to memory but have always been an adjunct to a large number of more commonly known facts already present in the brain. The potential problem lies in how an increasing reliance on the Internet might erode the line between facts that we can assume almost everyone knows and the kinds of facts that may well not be general knowledge and that you'd always need to look up. For example, if two adults in the developed Western world met each other today, they could take it for granted that they would both know what and where Barcelona is, or who Napoleon or Shakespeare was, without having to look it up on their mobile phone. They would be able to have an interesting conversation on the assumption that they shared a sufficient number of certain basic facts, a common conceptual framework that provides a starting point for developing ideas. What we have in common with others already largely determines the scope of our interaction and conversation, but let's take it to an extreme. Imagine that in the future people become so used to external access for any form of reference that they have not internalized any facts at all, let alone put them into a context to appreciate their significance and to understand them. Any discussion would be punctuated by lengthy pauses while each interlocutor looked up a name or a phrase on a digital device. Of course, some people have always known more than others. There has never been a clear divide as to what we can assume everyone knows and what is considered arcane and therefore acceptable to admit ignorance of. But if the balance eventually shifts more in one direction, perhaps normal, real-time face-to-face conversation (already imperiled by social networking sites) may be downgraded to the simplest interchanges where minimal general knowledge is assumed, or slowed down to such an extent that offline conversations, via text or email, become even more the norm.

The ease of looking something up on a search engine is already transforming not just memory strategies but our thought processes themselves. It is hard now to think back to the days of the question-rich, answer-poor environment in which many of us were students, a world where we had to leaf through heavy and cumbersome encyclopedias or plan a time-consuming trip to a reference library. Nothing came quickly or easily: there was a constant uphill struggle to obtain the exact information you wanted, and you had to focus on what was really essential. When you try to find an answer to a question, you're on a quest, a journey with a very clear goal: each step is sequentially linked in a linear path that eventually leads to a specific and different destination. As we've seen, this is how a thought process would differ from a raw instantaneous feeling, through the sense of a narrative over time. It is this experience of a goal-directed passage of time that I've suggested gives each of us a unique life story and the events and people within it a unique meaning. As T. S. Eliot so eloquently described it in *Little Gidding*:

We shall not cease from exploration
And the end of all our exploring
Will be to arrive where we started
And know the place for the first time.[8]

This last line is the whole point: the original place is actually now somewhere different. The very effort we invest in the journey of discovery, in the time spent joining the dots and making connections across networks of neurons, gives an importance, a significance, to what we learn, so we see things in a new way. Now we are in danger of entering the reverse scenario, an arguably question-poor world where our brains are saturation-bombed with answers but where it is hard not to be distracted and lose sight of what we wanted to know at the outset.

James Thurber, the American author, cartoonist, and celebrated wit, who died in 1961, well before "surfing" ever meant anything other than surmounting looming walls of salty water, once said: "It is better to ask some of the questions than to know all of the answers."[9] The experience of endless surfing over an infinite sea of responses might trump the original goal of articulating a question in order to find a defined and definite answer. In turn, this new easy-come, easy-go way of handling incoming information may have new effects on the ever adaptable human brain. In order to investigate this possibility, we need to unpack what may be happening to the mind when it is inundated by so much content. It's not just the amount of material available but, perhaps more crucially, the speed and therefore ease with which we can all interact and deal with it.

Now, thanks to Google and other search engines, we've gone from articulating questions to weaving and bobbing through answers. The Internet presents an endless stream of facts, but deep and interesting questions remain less obvious. Consider the example in the Sparrow study mentioned above: "An ostrich's eye is bigger than its brain." It may be that you never set out to learn much about ostriches, but in the course of Googling, say, "eyes," this fact popped up. In and of itself, the fact won't help you understand how eyes work, if indeed that had been your original question, but it will distract you, make you pause for a moment to say "Wow," and then get stored away in your memory as an isolated, disconnected fact that you might pull out when conversation in the bar or at the water cooler lapses. At best it will rupture a linear train of exploration in finding out about eyes, and at worst it will have confused you as to what the most important issues concerning eyes might be.

The problem could now be one not so much of relying too heavily on an external source for facts but of letting that mentality of collecting isolated bits and pieces of information overtake the formerly normal process of making use of these facts, of joining up the dots, as would normally happen in an internalized conceptual framework. In a 2013 investigation by Malinda Desjarlais at Brock University, undergraduate students with high and low levels of sustained attention were tasked with navigating the Internet for twenty minutes to learn about how tropical cyclones form, a topic about which they knew little at the outset; this was then followed by a test.[10] Students with high levels of sustained attention more frequently guided their learning in a linear manner, alternating between search engine results and first links. Rarely did these learners select hyperlinks presented on the links themselves, and it was these students who performed better on the test. Learners with low levels of sustained attention typically took advantage of the opportunity to jump around between sources of information. While they alternated between search engine results and first links, the low-attending learners engaged in far more exploration of the hyperlinks presented than the high-attending learners did. However, the hyperlinked sources were typically irrelevant. So, perhaps not surprisingly, those with short attention spans performed more poorly on the test than those who were able to focus longer.

Such variations in performance can be even more marked when we look across age groups. David Nicholas, director of CIBER Research, has investigated how different generations use the Internet to search for information, and the confidence they have in their search abilities. The Google generation (born after 1993), Generation Y (born after 1973 and before 1994) and Generation X (born 1973 or earlier) were compared for their Internet information-seeking abilities. The younger generations spent a fraction of the time the older generation did looking for an answer to both simple and complex questions. However, by their own admission, they were less confident about the answers they found, as demonstrated by the fact that they viewed fewer pages, visited fewer domains, and undertook fewer searches compared to the older group. Also, tellingly, the answers they provided to the simple and complex problems were much more the product of cut and paste. The younger generation also turned out to have poorer working memory and to be less competent at multitasking, despite engaging in it more. The researchers came to the conclusion that "the propensity to rush, rely on point-and-click, first-up-on-Google answers, along with growing unwillingness to wrestle with nuances or uncertainties or inability to evaluate information, keeps the young especially stuck on the surface of the 'information' age, too often sacrificing depth for breadth."[11]

These findings have profound implications for the Digital Natives and their ability to research information on the Internet, and more widely still for learning in general and thus overall success in life. Those with more facts at their immediate disposal can build richer constructs of reality and thus have a worldview informed by a context that enables deeper understanding—more *wisdom.* Although the number of facts internalized does not automatically guarantee wisdom, facts constitute the all-important dots that you connect, interpret, and place within your personal schemas to give them meaning. But if you can only remember the places to look for answers rather than the answers themselves, then even these dots will not be learned and therefore cannot be joined up with other dots to form an individual perspective of the world.

Another unprecedented experience offered by search engines

that could impact on how and what we learn is YouTube.[12] Watching videos on YouTube or similar sites is a form of learning in the most general sense, since watching a video involves the processing of an input coming into your brain from the screen. After all, you have acquired a tiny nugget of information; you now know something that you didn't know previously, even if it be that a cycling dog is alive and well and performing in Ohio. But many people watch YouTube videos without any explicit motivation to acquire any new information. The appeal is that YouTube presents visual information, actions rather than spoken words. Actions do indeed speak louder than words, and the watching of actions rare, exciting, or funny anchors you in the moment, as what you see is what you get. Granted, YouTube also enables commentary, and links are frequently shared between friends, so social networking can also flourish around a video, just as it might over a film or a book. The big difference is that because a video clip is usually limited to fifteen minutes or so, unlike a film or book, the YouTube video typically has a shorter and therefore less complex story to tell.

An action such as a dog cycling or humans dancing the Harlem Shake (where different groups of people dance to a song of the same name) has a face value all of its own; it need not stand for or symbolize anything unless it is placed in an elaborate conceptual framework of a story where the behavior has associations with previous actions or specific characters that give it a special relevance not intrinsic to the physical features of the event. It is very rare for such elaborate or complex story lines to be played out on YouTube; by contrast, television is more hospitable to such stories. Yet, while there is some evidence of substitution of Web viewing for conventional television viewing, the time spent viewing programming on the Web—on average some 6.8 hours per week—far exceeds the reduction in weekly traditional television viewing, which is only seven minutes.[13] Perhaps more important still, overall time spent on network-controlled viewing (television plus network websites) has increased by almost four hours a week.

In real life, actions always have consequences and, as we know only too well, cannot be reversed. Unlike in videogames, no one can become undead; killing someone is therefore a highly significant and meaningful act. By contrast, as we've discussed, dropping something on the floor and immediately picking it up is meaningless: the action has effectively been completely reversed. Most of life, however, unfolds between these two extremes: much of what we do seems meaningless at the time, but on reflection we realize that it set in motion a chain of cause-and-effect reactions that gave rise to a certain outcome. Even the dropping of a coin and its retrieval may lead to a particular outcome, even if only that people watching from now on regard you as a bit weird.

Alternatively, actions may lead not just to a predictable immediate effect but to one with many indirect ramifications. Surely it is this intricate sequencing of cause and effect, of indirect consequences, that amounts to a good story. The more unpredictable (but in retrospect understandable) the sequence of cause and effect—say, in a murder mystery—the more absorbing the narrative. If, on top of that, the characters also have intrinsic significance by virtue of what they've done in the past or simply because of their association with other characters, then the story is even better still: it is just like real life. By contrast, a character in a YouTube clip usually has no complex backstory and no personal relationships, and his or her actions have no long-term consequences; they are frozen in a tiny window of time. What you are watching doesn't really *mean* anything.

Could such a statement also apply to the freeze-frame of a painting? No, because a painting is showing you the world through the highly subjective and idiosyncratic eye of the artist, perhaps prompting new ideas and perspectives. If anything, a better analogy would be a photo or a series of photos of people, objects, and events with which you have no connection. Given the millions of videos that YouTube hosts, perhaps the competition between them for your attention, and the ease and speed with which they can be circulated, may suggest that quantity trumps quality, and that brevity is interlinked with a shorter attention span and hence a lower level of personal involvement or insight.

Therefore, it may seem baffling, sad, worrying, or to some perfectly understandable that people wish to spend their time passively watching something that is not necessarily even a story but makes you smile, gasp, shake your head, or cry, just for a moment. This is perhaps the most minimal activity of all those associated with digital technologies: for a few moments the outside world is replaced by the cyber one, for no purpose, requiring no response, making no point other than capturing your passive attention briefly. Then, of course, you can play it back again and again.

Perhaps it is the time-out from real life—the fact that no effort, no input, and arguably not even any thought is required—that is the appeal. If so, we have come a long way from both committing facts to memory and learning so that we can translate information into knowledge.

> What I . . . found fascinating [about asking people where they did their best thinking] was that only one person said in the office, and they said very early in the morning . . . in other words, when the building wasn't really functioning as an office at all. Interestingly, not a single person mentioned digital technology. . . . Technology, it seems, is good for spreading and developing ideas, but not much use for hatching them.[14]

Once again, futurist Richard Watson is the pessimist. But as our society spends increasing amounts of time surfing, swimming, or drowning in Google or YouTube, perhaps Watson has a point. The magical *something* about surfing may not be the value of infinite content, unprecedented speed, and ease of access. Perhaps the opportunity for an experience that can be an end in itself and that is impossible to obtain elsewhere is the true appeal. This online experience could easily trump the longer-term reason for surfing in the first place: to find something out. If so, then we are about to witness a radical change in how the next generation thinks.

THE SCREEN IS
THE MESSAGE

Back in 1964 Marshall McLuhan argued in his now legendary work, *Understanding Media,* that technology wasn't a neutral conduit but that in and of itself it would have an impact on mental processes: "The medium is the message."[1] McLuhan then went on to develop the distinction between "hot" and "cold" media. "Hot" media does most of the work for you; with TV, radio, or even a simple photograph, you are nothing but a passive recipient. In contrast, "cold" media, such as a cartoon or a telephone, require some kind of input from you, in response to a much more minimal offering. Interestingly enough, cyberexperiences can be regarded as both hot, because their ever more exotic and startling screen displays leave nothing to the imagination, and yet also cold, since their huge appeal comes from the interactive, participative experience they offer. The very medium of the digital technologies, the screen itself and what lies behind it, might now be driving our thought processes in an unprecedented direction. The physical difference between a screen and a book, the availability of hypertext, and the opportunity to multitask or engage in brain-training regimes are all unprecedented in their possible impact on our brain processes.

The first, most obvious physical feature of the screen is that the text is lit up on a hard surface as opposed to being printed on a fragile page. Back in 2001, Abigail Sellen and Richard Harper argued in *The Myth of the Paperless Office* that good old-fashioned paper would continue to play an important role in office life.[2] The basis of their rationale was the fascinating concept of *affordances,* the idea that the physical properties of an object "afford," or allow, certain activities. The reasoning was that paper, which can be thin, light, porous, and opaque, affords activities such as grasping, carrying, folding, writing, and so on. The affordances of the laptop and mobile phone will be very different.

Anne Mangen at the Oslo and Akershus University College of Applied Sciences set out to explore the importance of the affordance of actually touching paper, by comparing the performances of readers of paper with those who read on the screen.[3] Her conclusion was that e-reading resulted in poorer comprehension, as a result of the physical limitations of the text that forced readers to scroll up and down, thereby disrupting their reading with a spatial instability.[4] This is an important factor, since having a good spatial mental representation of the physical layout of the text leads to better reading comprehension. Those who understand well, compared to those who comprehend poorly, are significantly better at remembering and relocating the spatial order of information in a text, so there could well be a link between the physical layout of what you're reading and how well you understand it.[5]

A further consideration in reading from a screen is the greater potential for eyestrain. The differences between the printed page and the screen have significant consequences for visual ergonomics. The visuo-spatial perceptual processes of reading rely on the legibility of the text, which in turn is dependent on letter detection and word identification, light source, ambient luminance, character size, display time, interline spacing, and so on. Each of these processes impacts reading performance, visual fatigue, and search time. Even between different types of electronic media, lighting is a differentiating factor.[6]

Hanho Jeong, from Chongshin University, Seoul, aimed to assess the usability of electronic and paper books with objective measures such as eye fatigue, along with perception and reading comprehension in sixth-year state school students.[7] The results showed a significant "book effect" on quiz scores: compared to reading eBooks, reading paper books resulted in better reading comprehension. Moreover, the students had significantly greater eye fatigue after reading eBooks than after reading the paper counterparts, and although they were "satisfied" with the eBook, they claimed they actually preferred paper books. Most of them grew tired of reading on the screen. In turn, this fatigue could have an adverse effect on both reading comprehension and the perception of eBooks: further analysis of users' responses showed that many of their critical remarks were based on the screen or text size or clarity, rather than on the eBook itself.

A second distinguishing feature of digital technology is the temptation and opportunity it offers for multitasking. In his book *The Shallows,* Nicholas Carr is in no doubt as to the potential detrimental effects: "The Net seizes our attention only to scatter it. We focus intensively on the medium itself, on the flickering screen, but we're distracted by the medium's rapid-fire delivery of competing messages and stimuli."[8] Media multitasking can be operationally defined by all too familiar and highly irritating scenarios such as switching from checking emails to having an instant messaging conversation with someone, text messaging while watching television, or jumping from one website to another. In a survey of two thousand children between the ages of eight and eighteen, the time spent multitasking between more than one technology medium in 1999 was 16 percent but had almost doubled to 29 percent ten years later.[9] In a survey of U.S. college students, 38 percent said they could not go for more than ten minutes while studying without checking their laptop, smartphone, tablet, or e-reader.[10]

Since media multitasking involves, by definition, shifting attention between multiple sources, much research has focused on how much information can be retained, and how efficiently, when individuals multitask between media. One study put students through a series of three tests. In each case, the subjects were split into two groups: those who regularly engaged in frequent media multitasking and those who didn't. The three tests in the study involved the subjects looking at shapes, numbers, or letters, but the task was to remember something about just some of the images on the screen and to ignore the others.

In all three tests, the high multitaskers seemed unable to ignore the shapes they were told to ignore, and were unable to filter out what wasn't important to that particular task. In all cases the low multitaskers outperformed their high-multitasking counterparts. The researchers had originally set out to learn what benefits multitasking conferred, but Eyal Ophir, the study's lead author and a researcher at Stanford's Communication Between Humans and Interactive Multimedia Lab, concluded: "We kept looking for what they're better at and we didn't find it." Ophir's explanation was that "the high multi-taskers are always drawing from all the information in front of them. They can't keep things separate in their minds."[11] Anthony Wagner, a psychologist, amplified this idea further: "When they [high multitaskers] are in situations where there are multiple sources of information coming from the external world or emerging out of memory, they're not able to filter out what's not relevant to their current goal. That failure to filter means they're slowed down by that irrelevant information."[12]

Multitasking has also been cited as a reason for why the reading time for an e-textbook is significantly longer than for a paper book.[13] Research also shows that college students multitask for approximately 42 percent of class time.[14] An experimental investigation into multitasking and lecture comprehension found that comprehension was significantly impaired when students were assigned simple Google, YouTube, or Facebook search tasks that occupied only 33 percent of class time.[15] Overall, students who multitasked for a third of the lecture scored 11 percent lower on a post-lecture comprehension test. One answer to this apparent sad state of affairs is fairly simple: students who want to learn will do so, and those who become bored or unmotivated during lectures switch off. However, investigators went one step further and found that, for students who were not multitasking themselves, even the visible presence of other students multitasking during a lecture had a significant negative effect on their comprehension. Students who were in direct view of a multitasking student perusing Facebook, Google, or YouTube had a 17 percent poorer performance on the subsequent comprehension test, indicating that the distracting effect of personal computer technologies in the classroom had an impact not just on bored students but also on motivated ones.

Outside of class, is multitasking during study periods also affecting academic performance? Researchers observed middle school, high school, and university students engaged in academic work for just fifteen minutes in their homes.[16] They factored in the presence of other technologies and open computer windows in the learning environment prior to studying, and conducted a minute-by-minute assessment of on-task behavior and off-task technology use. Astonishingly, students averaged *less than six minutes* on task prior to switching, most often as a result of technological distractions (including social media and texting) and a self-reported preference for task switching. Having a positive attitude toward technology did not affect being on task during studying; however, those who preferred to task-switch had more distracting technologies available and were more likely to be off task than others. It's no real surprise that concentration is the key and that multitasking can be counterproductive.

Instant messaging has become one of the most popular forms of computer-mediated communication for college students, through programs such as Skype and Facebook Chat. Unsurprisingly, in a large-sample Web-based survey of college students, more than half of them reported that instant messaging while studying had a detrimental effect on their academic performance.[17] Similarly, two studies have found that there is a negative relationship between amount of time spent on Facebook and grade point average.[18] Facebook users also reported spending fewer hours each week studying compared to nonusers.[19]

While students may be aware of the detrimental effect of multitasking, a more formal investigation set out to measure how well students can perform on a test when they have been multitasking during study. In one investigation, the prediction was that students who engaged in instant messaging while reading a psychology passage online would take longer to read the passage and would perform more poorly in a comprehension test.[20] Participants were randomly assigned to one of three conditions: instant messaging before reading, instant messaging during reading, or no instant messaging. Students took significantly longer to read the passage when they were instant messaging during reading, not including the time taken actually to send the message. The researchers cautioned that students might feel as though they are achieving more in a shorter period of time while multitasking, which was patently not the case.[21]

In summary, although the ability to be in engaged in several things at once sounds like it might be wonderful for keeping pace with the speed of twenty-first-century life, the price paid could be high. Evidence is mounting regarding the negative effects of attempting to process different streams of information simultaneously, and results now indicate that multitasking leads to an increase in the time needed to achieve the same level of learning, as well as an increase in mistakes while processing information, compared to those who sequentially or serially process the same information.

A third basic feature of the screen that the printed book can never offer is hypertext. Although individual differences between readers, such as working memory capacity and background knowledge, all play a part in final reading performance, increased demands of hypertext on decision making and visual processing can have a detrimental effect on students' efficiency.[22] Hypertext is, after all, a deviation from the path of linear thought, a tangent that may or may not be a red herring, but you only discover which one it is once you've deviated. A hypertext detour that might lead to further meandering away from your initial intellectual journey arguably presents more of a distraction from the path of linear thought than, say, a traditional footnote, which is finite and leads no further. Moreover, a hypertext connection is not one that you have made yourself, and it will not necessarily have a place in your own unique line of reasoning and eventual conceptual framework. It will therefore not necessarily help you read at a pace that allows you to understand and digest what you're reading.

This notion of reading at your own pace is an important part of what is known as metacognition, or the ability to monitor and be aware of your own cognitive performance. Metacognition matches up closely with good reading comprehension. Rakefet Ackerman and Morris Goldsmith from the Technion–Israel Institute of Technology and the University of Haifa compared reading performance from on-screen learning and paper learning and found that performance did not differ significantly under fixed test conditions. However, when study time was self-regulated, screen reading produced a poorer performance than paper reading. The lower performance of those working from the screen was accompanied by significant overconfidence with regard to predicted performance, whereas subjects learning from paper monitored their performance more accurately. Ackerman and Goldsmith came to the conclusion that people appear to perceive the medium of print as more suitable for effortful learning, whereas the electronic medium, in this case a computer, is better suited for "fast and shallow reading of short texts such as news, emails, and forum notes . . . The common perception of screen presentation as an information source intended for shallow messages may reduce the mobilization of cognitive resources that is needed for effective self-regulation."[23]

This brings us to the fourth and most crucial issue of all: the reason for picking up a book or switching on an eBook in the first place. Recent research analyzing reading behavior in the digital environment over the past ten years has revealed that decreases in sustained attention are increasingly characterizing people's literacy skills and habits.[24] With a growing amount of time spent reading electronic documents, a profile of screen-based reading behavior is emerging characterized by more time spent browsing and scanning, keyword spotting, one-time reading, nonlinear reading, and reading more selectively, while less time is spent on in-depth and concentrated reading. So reading on a screen may both take longer than reading a book (because of the potential for distractions, such as hypertext links) and encourage a more browsing-oriented strategy. Which of the two, the book or the screen, might be the harder work?

At Johannes Gutenberg University in Germany, Franziska Kretzschmar's team measured brain waves (via EEG) and eye tracking to evaluate the cognitive effort involved in reading in each type of medium.[25] Results replicated previous findings in that participants overwhelmingly chose the paper page over an e-reader or a tablet computer as their preferred reading vehicle. However, actual cognitive effort did not differ between media, indicating that while readers subjectively rank digital devices as requiring more of an effort, objective results in terms of comprehension or cognition were indistinguishable. This subjective perception may account for why electronic textbooks are still not very popular with college students. Textbooks will be read for different reasons and with different strategies than, say, novels.[26]

Certainly skills beyond comprehension and cognition may flourish more readily as a result of reading paper textbooks. For example, one investigation at Sheffield University followed students as they identified woodlice, with one group using a conventional paper-based identification guide, and the other group using the same key on a computer.[27] It turned out that the group using conventional textbooks was more curious and questioning of the information. Perhaps a book has a sense of permanence and immediate structure that enables students to feel more secure and confident in asking questions. Alternatively, they may feel that they have more time to reflect, that there is no rush to press a key for the next entry on the screen. So perhaps it's this sense of personal exploration at their own pace that underlies the subjective preference of students, as seen in the other studies.

Yet here's the paradox: despite the appeal of paper books, reading is becoming an increasingly digital experience. EBook sales are rapidly increasing, while sales for traditional books have slowed down.[28] In the United States in 2012, eBook sales outpaced hardcover book sales for the first time.[29] The slow growth in the sale of paper books will inevitably have consequences for retailers. Independent booksellers in the United Kingdom have been shutting down one after another; their numbers are now down a third since 2005.[30] Overriding socioeconomic and lifestyle factors such as the novelty, cheapness, and accessibility of eBooks are clearly key factors that are trumping other, more intellectual considerations. Books and screens offer very different kinds of experiences and consequently will elicit different performances, responses, and priorities.

Perhaps the greater appeal of the printed book, but one that will not be at a premium when the more workaday considerations of money and convenience are taken into account, is its cultural symbolism. Paper books are of a fixed time and place, and their permanence offers a reassuring security that an eBook can never deliver. As I look around my study, with bookshelves covering three of the four walls around me, I try to imagine these walls denuded in favor of a small stash of flash drives. Just seeing and touching the books—

some in hardcover, some in paperback, in different colors and sizes, and in varying degrees of dilapidation—is like being surrounded by old friends. In many cases I remember the time in my life when I acquired a certain book and devoured the facts it contained, or was stunned by the ideas it set out. Even though the contents of some of them may have become obsolete, the prospect of throwing out any one of these books, or indeed any book at all, would seem almost like a kind of murder.

Beyond the functional value offered by the particular properties of the printed page, and beyond personal memories, there is also the powerful iconography of physical books. On May 10, 1933, Nazi students burned upward of twenty-five thousand volumes of "un-German" books, including the writings of Albert Einstein as well as those of non-German authors such as Ernest Hemingway. On that site in Berlin today, a large glass-covered opening in the cobbled square reveals an excavated area below of wide walls of empty bookshelves in a simple but chilling testimony . . . to what? Books stand for knowledge, new ideas, and the inventiveness of the human spirit and imagination. Will the Digital Native in the future appreciate the inherent value of such a non-interactive object with its fixed time and place, its unchangeable story locked away in its fragile pages?

Printed works may always have something special to offer, despite our changing lifestyle, agenda, and mindset. Books and screens may become complementary objects rather than rivals, just as the book and the movie, the radio and the TV, or the bicycle and the car play different but complementary roles in many of our lives. A new part of that life now is the acquisition of facts through digital devices. Will this shift in medium change how *effectively* we process those facts—how we learn, remember, and think?

Aside from the more general activities of hypertext and multitasking, digital technologies could offer unique formal pedagogic opportunities. There are many brain-training products that claim to improve cognitive function through the regular use of screen-based exercises, and modest but positive effects from their use have been reported in some studies of older individuals and of preschool children.[31] Adrian Owen and his colleagues in Cambridge and London were nonetheless not convinced there was sufficiently hard empirical evidence of their efficacy.[32] They investigated the key question of whether the benefits accrued during training would transfer to other untrained tasks, or indeed lead to any general improvement in the level of cognitive functioning. During a six-week online study, they monitored some eleven thousand participants trained several times each week in cognitive tasks designed to improve reasoning, memory, planning, visuo-spatial skills, and attention. Improvements were apparent in every one of the cognitive tasks in question, as might have been expected, but the crucial observation was that there was no evidence for the transfer of these effects to untrained tasks, even when those tasks were closely related in terms of the thought processes required.

But wait a moment. Didn't we just see in the earlier discussion on videogames the precise opposite—that there was indeed robust evidence that the skills learned while gaming could be transferred to more generalized contexts? So, simply at face value and without drilling down into the relative merits of specific games and training regimes, where might the crucial distinction lie? One important difference, with all the usual caveats of stereotyping, might be that videogames are by definition providing an experience that is more exciting and stimulating than what the dull, three-dimensional world can offer. In contrast, brain training is rarely marketed as exciting. After all, if we are thinking in terms of the serious acquisition of knowledge and selling that as a product, it is the long-term acquisition that needs to be emphasized to the customer rather than a short-term moment of frivolous fun. The motivation for buying a brain-training program is self-improvement. On the other hand, the primary reason for choosing to play a videogame is not to learn but to enjoy yourself.

The difference between short-term sensation and long-term cognitive improvement in the brain is, at least in part, determined once again by the participation of our faithful old friend, the neurotransmitter dopamine. Could the presence or absence of high dopamine levels make the difference, at least hypothetically, between whether or not you can apply a skill learned from one task to other tasks and activities? Simplistic though it might seem, one possibility springs from the fact that dopamine operates like a fountain in the brain, emanating from the more evolutionarily basic parts to access wide reaches of the "higher" cerebral regions. Dopamine can also serve as a modulator, acting as an agent that can predispose brain cells to be more sensitive to stimulation when it arrives. Scenarios such as gaming, where dopamine is released as a result of raised arousal and reward, could enable more brain circuitry to be harnessed, and hence for learning to be more generalized.

We shouldn't ever underestimate the importance of enjoyment. Part of the appeal in studying lies in its potential for social interaction, the feeling that it gives us that we belong, that we are part of the crowd and are not being left out. Networked interactivity is one of the essential factors that differentiate the most recent online educational games from traditional stand-alone CD-based games. Kwan Min Lee and his colleagues from the University of Southern California measured how networked interactivity influenced game users' learning outcomes in online educational quiz games, offline educational quiz games, and traditional classroom lectures.[33] The researchers found that networked interactivity in the online educational quiz games enhanced game users' positive evaluation of learning, test performance, and feelings of social presence. Further analyses indicated that it was the *feelings* of social presence in an interactive network that counted in the various learning outcomes. So by promoting the feeling of being connected with others, screen technologies act as a positive driver. It is unsurprising that the best environment for learning turns out to be one in which you are having fun and interacting with others, irrespective of whether these key ingredients are provided through a screen or a more traditional scenario.

While the screen can readily offer a more rigorous rehearsal regime in mental processing than people or paper ever can, does that mean we learn more *effectively* from a screen? Computer-assisted technologies have, of course, been employed in classrooms for decades, and moderate use of them continues to enhance students' learning experience. In particular, the case for screen devices in education seems most conclusive for special-needs students, whether they have a visual impairment, dyslexia, or some other learning difficulty. So far, the use of "errorless" software, where there are no right or wrong answers, has proved to be one of the best approaches. With this software, trial and error, as well as exploration, are rewarded with fun noises, humorous animations, vivid graphics, music, and natural-sounding speech. For children with special learning needs, this nonjudgmental interactive software, with its fast-paced and colorful displays, is easily more motivational than a simple printed book.[34]

Touch tablet devices certainly seem to be beneficial for a range of students with developmental disabilities. One review looked at fifteen studies covering five domains: academics, communication, employment, leisure, and transitioning across school settings.[35] The studies in question reported outcomes for participants who ranged from four to twenty-seven years of age and had a diagnosis of autism spectrum disorder and/or intellectual disability. Most studies involved the use of iPods or iPads and aimed either to deliver instructional prompts via the device or to teach the individual to operate the device to access preferred stimuli. The latter goal also included operating the device to generate speech as a means of requesting preferred stimuli. Taken together, the results were largely positive, suggesting that iPod, iPod Touch, iPad, and related devices are viable technological aids for individuals with developmental disabilities.

The benefits of screen technology are also evident in mainstream learning. For example, one meta-analysis of forty-six different original studies involving a total of 36,793 students showed significant positive effects of computer use on mathematics achievement.[36] Similarly, a recent large-scale analysis reviewed how educational software programs affect reading outcomes in a total of eighty-four studies involving more than sixty thousand students.[37] The findings suggested that various reading programs, predominantly computer delivered, generally produced a positive, though small, effect on reading skills. However, any innovative technology application or integrated literacy intervention showed more positive results when there was teacher support. So the greatest promise of the digital devices lies not so much in the software and screen delivery themselves, but in their use in close connection with teachers' efforts.

For anyone who's read *The Prime of Miss Jean Brodie* or *Goodbye Mr. Chips,* this will come as no new insight. Nothing beats an inspirational and exciting teacher. However, direct face-to-face instruction is declining in higher education. Lecturers have also observed another trend in university courses: 55 percent of academic staff recently reported that lecture attendance had decreased as a result of introducing digital audio recordings of their presentations.[38] Back in 2006, one of the main reasons college students gave for not attending lectures was the availability of materials online.[39] In the same spirit, when asked why they did not attend lectures, almost 70 percent of students surveyed at an elite Australian university reported that they could learn as effectively using digital audio recordings as they could by attending the corresponding lecture in person.

However, one report found that economics students who learned course material via virtual delivery performed significantly worse compared to those who attended traditional lectures.[40] While the two groups did not differ in regard to their grasp of basic concepts, the group that learned virtually fell significantly behind in their grasp of complex material. This indicates that sophisticated ideas cannot be transferred via the screen as effectively as in person. Another study found similar results favoring person-to-person instruction for academic performance.[41] Indeed, when college students in a large introductory microeconomics course were randomly assigned either face-to-face lectures or video-streamed presentations, the students who attended face-to-face lectures had higher average test scores.

It seems that the benefits of dialogue, face-to-face discussions of issues, and problem solving with another person still exceed the benefits of virtual communication. When it comes to education, surely there is always a strong case for real classrooms with real teachers overseeing real-life conversations, regardless of the number of screens in a classroom and the time spent in front of them. Recent studies suggest that eBooks and tablets might be useful educational tools—but, crucially, only when used alongside adult supervision. Ofra Korat and Adina Shamir at the School of Education, Bar-Ilan University, examined the effects of eBook reading on the reading skills of children ages five and six.[42] While one group read an eBook independently, a second group read an eBook with adult mediation, a third read the printed book with adult mediation, and the fourth read the printed book with no adult intervention. The results showed that the activity of reading the eBook with adult assistance produced greater progress in the recognition of letter names, emergent word reading, and general emergent reading level than all other groups. Here the eBook might be superior to a traditional book, *provided an adult is around.*

Education doesn't take place in a bubble but is an integral part of a person's life and relationships. Different lifestyles will therefore also play their part in determining whether and to what extent the screen can make a difference to learning. For example, another factor associated with higher test scores in mathematics and reading is having a home computer, even after allowing for family income and for cultural and social capital.[43] However, home computing may generate a "Sesame Street effect," whereby an innovation that held great promise for allowing poorer children to catch up with more affluent children educationally actually widens the educational gap between affluent and poor, between boys and girls, and between ethnic minorities and whites. This gap could grow as different must-have (i.e., expensive) digital devices appear at an ever faster rate.

The iPad is now a mainstay of education and entertainment for many children. While most schools in the United States don't have the purchasing power to provide all their students with iPads, the children who *do* have them are getting them from their families and other adults, who presumably are using them as well. The iPad plays an increasingly important part in the American educational system. In a recent list of the top one hundred enterprises with the largest iPad roll-outs worldwide, nearly 70 percent of the list were U.S. schools.[44] In 2013 Apple signed a $30 million deal with the Los Angeles Unified School District, the second-largest public school district in the United States, to provide every student with an iPad by 2014.[45] Other Western countries are also zealously integrating iPad technology into the formal education system.

Schools around the world are adopting tablet-only classrooms (known as "one-to-one classrooms" and fully supported by Apple, for whom the commercial implications shouldn't be ignored) for students as young as kindergarten age. One elementary school in Arizona outfitted a classroom solely with iPads, dubbing it the "iMaginarium."[46] If we're going to try to evaluate how the newly pervasive cyberculture affects how the brain adapts to different styles of learning, the large-scale introduction of iPads into the classroom might be a good place to start.

Consider, for example, an email I received recently from a concerned mother who is also a physician:

My daughter's school in Australia is aggressively introducing digital learning from grade five. . . . They will use nothing but a computer slate from the age of nine or ten which is also Web enabled. As a health professional, I have done extensive Internet searches myself and am yet to find any evidence for the benefits apart from "expert opinion" and anecdotes. Do you know of any scientific evidence for the neurophysiological effects of using nothing else but computers for learning?

One typical iPad enthusiast is Lisa Wright, head of a school in Essex in the United Kingdom, who claims the flexibility of the curriculum means that iPads could be used right across the primary school. Wright is a clear convert:

Year Four children [eight or nine years old] have used them in maths lessons and reception children have played some maths and phonics games . . . Year Ones [four- or five-year-olds] had them in their religious education lesson and Year Five and Six pupils [nine- to eleven-year-olds] have been using iPads in their topics, such as learning about the *Titanic* by getting on the Internet . . . We bought the iPads because they're so flexible and versatile. We've got a lovely outdoor space here so the children can take them outside and even use them to take pictures. We want learning to be fun for the children. The iPads are in use all the time. If you walk around the school, there's a child somewhere or a group using the iPad, which is what I want to see.[47]

Although Mrs. Wright also insists that books and conventional teaching methods, such as pencil and paper, are equally important, in many one-to-one classrooms the tablet computer has replaced all traditional teaching methods.

In contrast to such an overwhelming vote of confidence, a recent report claimed that millions of dollars' worth of tablet computers were sitting in British school cupboards as a result of teachers' overenthusiasm in purchasing new technology without any evidence that it actually improved educational outcomes.[48] We often assume that any new technology is automatically superior to what has come before; advances in knowledge and understanding are ascribed to the gadget itself. This view is frequently based on availability and novelty but not on other factors such as the type of supervision being given or the teacher's ability to inspire students. More to the point, though, to take up the question asked of me by the Australian mother-physician, just what evidence is there that iPads and other digital aids really do make a serious difference?

A critical and potentially confounding factor to bear in mind is the formidable physical appeal of the iPhone and iPad. David Furió and his team at the Polytechnic University of Valencia set out to compare learning outcomes and preferences of children eight to ten years old who played an educational game either in its traditional form or on an iPhone.[49] Ninety-six percent of the children indicated that they would like to play the iPhone game again, and 90 percent indicated that they preferred the experience with the iPhone game over the traditional one. The design of the physical object itself was clearly an important factor.

A similar result emerged in a 2013 study comparing desktop computers and iPads.[50] Students received an online multimedia lesson either on a desktop iMac in a lab or on an iPad in a courtyard outside. The students then experienced either a standard continuous lesson with no headings or an enhanced lesson where each slide had a helpful heading and where the learner clicked on a button to go on to the next slide. In both cases, perhaps not surprisingly, the group receiving the enhanced lesson outperformed the group receiving the standard lesson. However, regardless of the type of lesson they received, the iPad group rated themselves more willing to continue learning than the iMac group. Given that switching to iPad-based classrooms blindly assumes that traditional teaching materials are inferior, this current trend is very worrying. Until we have solid scientific evidence that iPads really do have superior pedagogic powers rather than just being prettier, it seems foolhardy to replace traditional teaching methods, which may actually be more effective, albeit less flashy, with these devices.

Interestingly, a backlash against the premature adoption of technology in classrooms is gathering momentum in California, with many schools opting for low-tech teaching methods. "Engagement is about human contact, the contact with the teacher, the contact with their peers," says a parent of three children who is also an employee of a high-tech company. Meanwhile, Paul Thomas, a former teacher and an associate professor of education at Furman University, who has written twelve books about public educational methods, stresses that "teaching is a human experience. Technology is a distraction when we need literacy, numeracy and critical thinking."[51]

In the United States there are 160 Waldorf schools, which subscribe to a teaching philosophy focused on physical activity and learning through creative, hands-on tasks. In fact, these schools ban all digital devices, as their credo is that computers inhibit creative thinking, movement, human interaction, and attention spans. Tellingly, the *New York Times* reported that the Waldorf school in Los Altos was popular with the very Silicon Valley parents who were themselves immersed in the digital industries.[52] This seems like a particularly fascinating trend, not only for education but for Mind Change as a whole. If the clever minds behind videogames, social networking, and tablets are wary about immersing their own children in these technologies, perhaps a general growing skepticism about their educational benefits is warranted.

One very extreme consequence of using high-tech methodologies in classrooms is the potential effect on literacy. If information is increasingly conveyed through the spoken word and visual images, we might have to face the possibility that literacy will be less and less relevant in our future lifestyle. Why learn to read or write when everyday communication can be so readily accomplished without either of these skills? Already literacy standards are declining: research has shown that many children are more likely to own a mobile phone than a book.[53] Another study by academics at Dundee University found that teenagers now prefer easier reads such as the Harry Potter and Twilight series.[54] Astonishingly, Eric Carle's classic picture-book *The Very Hungry Caterpillar,* which charts a caterpillar's transformation into a butterfly over a week, emerged as the most popular book among girls ages fourteen to sixteen.

The jury remains out on the value of pervasive digital technology in education; we will have to wait until the preteens of today take up their first jobs. Currently, it seems that any short- or medium-term impact will depend on the context in which screens feature: what is being taught, by whom, and where. More generally, for all of us, these powerful interactive screen technologies are not just exciting experiences but critical tools that have reshaped our cognitive processes and will continue to do so, creating both benefits and problems. The difference between silicon and paper, the distractions of multitasking and hypertext, and the tendency to browse rather than to think deeply all suggest fundamental shifts in how our brains are now being asked to work.

18

THINKING DIFFERENTLY

W hen the Nobel Prize–winning physicist Niels Bohr took his colleague to task for merely being logical instead of thinking, what particular talent in the human cognitive tool kit did the great intellectual pioneer feel was being neglected? Nothing less, presumably, than the quintessential mental activity that has enabled our species to probe into the meaning of our existence and to express those insights through science and the arts. Yet in today's digital culture, with its emphasis on computation, there's a danger that growing numbers of us are taking the more straightforward path and thinking increasingly like a computer, interacting with, and adapting to, its algorithmic mode of functioning.[1]

Sometimes such logical thinking is just what's required for solving a specified problem. Of course, problems come in all shapes and sizes, from simple IQ tests and Sudoku puzzles right up to resolving an economic crisis, trying to reignite the faltering Arab Spring, or coping with seemingly insoluble personal crises. But it's easiest to start with the most straightforward brain teasers, where, unlike real life, the problem has a clear and unambiguous solution. The skills needed here are the type of agile computational processing that is measured by IQ tests.[2]

Though many admit that intelligence can be defined and expressed in many ways, IT aficionados such as the physicist Ray Kurzweil focus on a narrow definition of intelligence, which they denote as *g,* and assume that this multifaceted phenomenon can be expressed as a computational process.[3] Contrary to popular belief, a high or low IQ may not be something you're simply born with. The largest genetic study on children has shown that only between 20 and 40 percent of *g* is inherited.[4] Leaving aside the question of how accurately IQ scores measure mental prowess, the strong impact of the environment can be evidenced in the significant and long-sustained increase in IQ scores seen in the past fifty to sixty years.[5] This increase, known as the "Flynn effect," may be caused by a number of factors, with the eponymous James Flynn himself suggesting that this rise may be due to the more stimulating environment of modern times.[6]

Another possible explanation for the rise in IQ test proficiency may be the increased rehearsal of test-specific skills. Since the beginning of the twentieth century, the explosive growth of films, television, videogames, and the Internet has exposed us to more visual media, allowing us to become increasingly adept at visual analysis. One variant of the IQ test, the Ravens Progressive Matrices IQ Test, emphasizes visuo-spatial skills; tellingly, the increase in those scores has been dramatic. Steven Johnson, author of *Everything Bad Is Good for You,* elaborates on the idea that gaming and competence at IQ tests exercise the same mental processes. As a result of increased interaction with the screen, Digital Natives are developing certain skills better than previous generations reared on books.[7] This suggestion seems persuasive when we compare the kinds of skills needed to perform well on IQ tests with those rehearsed in computer games. Both are abstracted processes, requiring the ability to see connections and anomalies and, above all, to detect rules independent of a wider context or any background knowledge. Johnson also suggests that screen culture is developing minds that are better adapted to greater complexity and have a greater proficiency at multitasking. This ability to solve problems while keeping in mind multiple rules and contingencies (working memory) is further enhanced by videogames that train us to solve problems faster or juggle problems at a faster rate.[8]

This is the most likely type of intelligence that our evolving cyberculture is helping to nurture, a computational ability that is already outstripped by silicon and which impresses Ray Kurzweil so much that he predicts that digital devices will one day supersede the human brain. However, Kurzweil overlooks the fact that computational processing requires a specific end point, a clear solution to a specified problem. The type of intelligence enhanced by prolonged screen interaction involves discerning patterns and processing connections so that the correct solution is reached within a given time. In contrast, other manifestations of intelligence, such as writing *War and Peace* or imagining how the brain might generate consciousness, are infinitely more open-ended. When the problem is finding the solution to a defined puzzle or searching for a fact, then accessing the screen will help. But if it is parsing the meaning of life, then juggling tasks and audiovisual expertise will be of little use.

"Game players are not soaking up moral counsel, life lessons or rich psychological portraits," Steven Johnson readily admits.[9] So what ability enables the human mind to progress beyond mere reasoning, to escape the computational mindset so admired by Kurzweil but cautioned against by Bohr?

Although IQ scores have risen, other abilities have remained constant. There has not been a concomitant increase in insights into the economic situation; no really noticeable increase in the creative arts; nor even on the horizons of neuroscience, compared to previous decades. However, it's important to bear in mind that the Flynn effect lies mainly in the middle range of ability, within the group of people who do not usually win a Nobel Prize, compose symphonies, or even just venture into politics, or the outreaches of academic research.

John Newton, head of Taunton School in Somerset, fears that "we will raise a generation who do not love learning but simply see the screen as a source of opinion or nuggets of information, poorly digested, that will suit their point of view without testing their veracity." Just as rote memorization differs from true learning, Newton believes, critical thinking requires "balance and a firm grasp of facts and context to avoid being led astray."[10] I have deliberately selected this quote from many similar ones voiced by teachers around the world because Newton highlights two crucial terms, "facts" and "context."

> Now, what I want is, Facts. Teach these boys and girls nothing but Facts. Facts alone are wanted in life. Plant nothing else, and root out everything else. You can only form the minds of reasoning animals upon Facts: nothing else will ever be of any service to them. This is the principle on which I bring up my own children, and this is the principle on which I bring up these children. Stick to Facts, sir.[11]

Extreme though it may seem, this view, expressed by Thomas Gradgrind in Charles Dickens's *Hard Times,* is perhaps closer to where the current mindset could be heading than we might care to admit.

> "Bitzer," said Thomas Gradgrind. "Your definition of a horse."
> "Quadruped. Graminivorous. Forty teeth, namely twenty-four grinders, four eye-teeth, and twelve incisive. Sheds coat in the spring; in marshy countries, sheds hoofs, too. Hoofs hard, but requiring to be shod with iron. Age known by marks in mouth."

The facts are indeed all there, and accurate. It's just that the dots are not joined up at any level, from the literal to the metaphorical. The Gradgrind approach conflates efficient information processing with real understanding: the insight and the knowledge that characterize a gifted mind involve more than the regurgitation of facts. You can train a brain (in certain cases even that of a parrot) to give the right responses to a given input, to recite poetry, or to answer factual questions with factual answers. But real intelligence requires a synthesis of facts, context, and meaning that encompasses far more than accurate responding.

Although we might access information efficiently and even regurgitate it on demand, success in such activities as Trivial Pursuit or bar quizzes is not regarded by even the most enthusiastic fans as the pinnacle of intellectual endeavor. Facts on their own are not enough! While collecting information is gathering dots, knowledge is joining them up, seeing one thing in terms of another and thereby understanding each component as part of a whole. The more connections you can make across an ever wider and more disparate range of knowledge, the more deeply you will understand something. Search engines and videogames do not provide that facility; nothing does, other than your own brain.

Even when you read at secondhand someone else's idea, whether in a book or in a condensed form on Google, it's only by incorporating it into *your* own personal conceptual framework that you derive your own take on whatever it may be. Hence *your* interpretation, *your* evaluation, *your* understanding will inevitably be individual to you and different from everyone else's. This conceptual framework is something that has been developing since you were small. Your experiences, the stories you hear from others and read yourself, and the facts you've been taught all build up into an ever more complex system of cross-referencing.[12] This connectivity, achieved through the plasticity of neuronal connections during development, may be the key feature that defines real learning, which sets the human brain above and beyond the information processing of a computer. This is why the concept of *context,* beyond mere facts, is so important.

When we seek to measure the kind of intelligence that comes into play when the problem we need to solve requires us to take the context into account—that is, when the question requires a "crystallized" intelligence, as discussed in Chapter 7—then the IQ score gains that we've been labeling the "Flynn effect" begin to evaporate. The Flynn effect is most visible on IQ tests that measure a more computational type of mental agility—the "fluid" intelligence discussed earlier.[13] Tests such as the Ravens, the Norwegian matrices, the Belgian Shapes, the Jenkins, and the Horn are all designed to measure fluid intelligence. They emphasize problem solving and minimize reliance on specific skills or familiarity with words and symbols. These are the tests that have shown an increase of about fifteen points per generation on average.[14] However, tests such as the Wechsler and the Stanford-Binet, which measure verbal abilities as well as more direct problem-solving skills, show fewer IQ gains, and would be less directly improved by a facility with videogames.

I've been suggesting that *meaning* is an association between at least two elements, whether they are objects, people, or events, or emotions. A wedding ring has particular significance if it is yours, even though it looks quite generic. The associations that this particular object, and no other, evokes imbue it with a special association for you that is not apparent to anyone else, nor is it intrinsic in the physical qualities of the ring.

So, the greater our ability to forge these links the greater our *understanding.* As we construct these associations, we bring together two previously disparate and independent elements, and can see one thing in terms of something else; for example, the snuffing out of a candle stands for the extinguishing of a life. As we live our individual existence, the linking of certain objects, people, and actions with previous objects, people, actions, and emotions will imbue them with a cognitive quality rather than a merely sensory quality, a meaning shared by no one else, which is unique to you. When we encounter a person or an object, we create personal meaning, and when we link that person or object to a wider framework our understanding grows richer and deeper. Finally, as we develop a sequence over time that links these meaningful things into a linear causal sequence, the original meaning and understanding changes and adapts. This is the kind of thought process that characterizes the mature human mind.

The work of the late educational neuroscientist John Geake provides hard evidence for this suggestion. Geake's imaging studies of gifted children revealed that their brains showed greater interconnectivity than the brains of those with average cognitive ability.[15] Specifically, the findings led to the idea that giftedness is linked to "analogical reasoning" (e.g., the analogy of the candle going out with death), a kind of reasoning that identifies, compares similarities between established concepts, and then uses those similarities to gain an understanding of new concepts. This ability to make connections where they didn't exist before, to connect the dots, could account for talents in a number of academic areas, including philosophy, mathematics, science, and music.[16]

A similar pattern seems to hold for adults. In Beijing, Professor Ming Song and his colleagues in the Chinese Academy of Sciences showed that brain imaging could demonstrate correlations between high intelligence and the strength of the functional connectivity distributed widely across the cortex.[17] The authors concluded that these observations were further evidence for a "network view of intelligence," and that this connectivity was operative even in the resting state and in the absence of any explicit cognitive tasks.

So if connections enable deeper understanding, then the process of making these connections can loosely be termed "thinking." In Chapter 1 I suggested that the crucial distinction between a raw feeling and a thought is a time frame. Simply being conscious, which is something any infant or nonhuman animal can achieve, always entails some kind of subjective sensation, as revealed by tail-wagging, purring, gurgling, or smiling. But never at any time does the animal in question suddenly turn into an automaton or zombie. It is impossible to disentangle consciousness from this subjective state of feeling. In fact, I would argue that they are pretty much synonymous.

By contrast, although all animals have degrees of consciousness, and therefore feeling, not all animals are capable of what we would recognize as thought processes. It is a skill that even humans have to develop as the years unfold. So what do a fantasy, a rational argument, a memory, a hope, a grievance, a business plan, and a joke all have in common? You start off in one place and end up somewhere else. And this sequence of linear steps unfolds over time, with a clear beginning, middle, and end. Unlike a raw feeling, the thought process transcends the here and now; it has to, as it links a past with a future.

In brain terms, the prefrontal cortex is once again pivotal. We've already seen how an underdeveloped prefrontal cortex is linked to an underdeveloped understanding both of figurative language as well as an ability to connect current actions to future consequences. It may not be surprising, therefore, that this part of the brain, when fully functional, plays a part in the human experience of time frames and time passing. Damage to the prefrontal cortex can, in addition to many other deficits, lead to "source amnesia"—not so much the loss of a memory as the loss of how and when a memory was created.[18] Memories will now be free-floating, no longer tethered to any personal context. If you have source amnesia, all your memories will blur together instead of being compartmentalized into specific incidents. You may remember a fact but not how and when you learned it. Your recollections would be more like the memories of a small child or a nonhuman animal, hazily aware of the past insofar as it colors the here and now but lacking any kind of order or chronology, and therefore any meaning. Your detailed life story will make no sense, not even to you.

The notion of the life story, or indeed any story, is compelling to most people, perhaps because it represents an amplification of the basic human thought process. The traditional custom of reading bedtime stories has been the best possible way to help children develop the cognitive skills of imagination, attention span, empathy, and insight into the minds of others. Research from the University at Buffalo, New York, measured the impact on the empathy of undergraduates reading passages from J. K. Rowling's Harry Potter books and Stephenie Meyer's Twilight series. Participants then answered questions designed to measure how they identified with the worlds they had been reading about. Results showed that participants who read the Harry Potter chapters self-identified as wizards, whereas participants who read the Twilight chapters self-identified as vampires. More fascinating still, membership in these fictional communities actually provided the same mood and life satisfaction people derive from affiliations with real-life groups. The authors of the study, Shira Gabriel and Ariana Young, concluded: "Books provide the opportunity for social connection and the blissful calm that comes from becoming part of something larger than oneself for a precious, fleeting moment."[19]

Although this particular study focused on college students, the power of stories and storytelling extends equally to adults. Keith Oatley, a professor in the Department of Human Development and Applied Psychology at the University of Toronto and a published novelist himself, expands on this point:

> I think the reason fiction but not non-fiction has the effect of improving empathy is because fiction is primarily about selves interacting with other selves in the social world. The subject matter of fiction is constantly about why she did this, or if that's the case what should he do now, and so on . . . In fiction, also, we are able to understand characters' actions from their interior point of view, by entering into their situations and minds, rather than the more exterior view of them that we usually have.[20]

A novel can, unsurprisingly, be as much of a learning tool as a textbook. We need fiction, someone else's story, in order to understand our own facts. The characters in question have a meaning because they can be linked in a conceptual framework, a context, to others and to past events, just like in our own lives. When we read fiction, as opposed to nonfiction, we are transported into the world of the characters and start to connect with them, the experiences they have, and the decisions they make. We may feel positive or negative emotions toward them as people and care deeply about what happens to them in a way that would be much less likely with a character in a videogame who is little more than an icon. The journalist Ben Macintyre sums it up beautifully:

> From the moment we become aware of others, we demand to be told stories that allow us to make sense of the world, to inhabit the mind of someone else. In old age we tell stories to make small museums of memory. It matters not whether the stories are true or imaginary. The narrative, whether oral or written, is a staple of every culture the world over. But stories demand time and concentration; the narrative does not simply transmit information, but invites the reader or listener to witness the unfolding of events.[21]

By observing what happens, by following the linear path of a story, we can convert information into knowledge in a way that emphasizing fast response and constant stimulation cannot. As I see it, the key issue is *narrative*. In a narrative there is a sequence—a chain of cause and effect in a nonrandom, strictly ordered sequence. Any narrative will, in some way, echo a life story. Stories arrange events into a context, a conceptual framework, and this order creates meaning. While narratives are the sine qua non of books, they are far from guaranteed on the Internet, where parallel choices, hypertexting, and randomized participation are more typical. While empathy may be developed from reading books, it may not be automatically guaranteed in a digital lifestyle that favors the rushed, the shallow, and the disconnected.

But surely search engines could free us up for more challenging questions and deeper thinking than we could ever have imagined possible, just as the printing press once granted more people access to knowledge. Maybe so, but we first need to have some story lines already in place. Without a personalized conceptual framework that enables us to use the Internet to frame and think about open-ended and difficult questions, we run the risk of being passively driven by isolated facts as we lurch from one isolated but amazing screen experience to another. As I mentioned earlier, it's worth noting that even the chair of Google, Eric Schmidt, believes that sitting down and reading a book "is the best way to really learn something."[22] We need time to think about and understand the world around us. The sequence of steps, the "movement confined to the brain," happens not in an instant but within a time frame as a train or line of thought. It seems that cyberculture does not encourage the development of the attention span necessary for deep thought, and thus if we rely exclusively on that digital culture, we fail to construct the adequate conceptual framework that gives the world around us meaning.

The reading of stories has to be the best possible way to develop the cognitive skills of imagination, attention span, and insight and empathy into the minds of others, as well as to provide us with a grasp of abstract concepts. After all, how would you convey honor, for instance, as an icon? Yet anyone reading Malory's *Morte d'Arthur* would get a sense of what honor means. Hence a novel can be as much of a learning tool as a textbook. We need fiction to understand facts. And if all this is so, then search engines are not the best vehicles for gaining understanding or for acquiring knowledge. The critical issue facing us is how to negotiate the transition from the old question-rich, answer-poor environment of the twentieth century to make sense of, indeed survive in, and make the most of, the current question-poor, answer-rich environment delivered by a fast-paced technology. In my view there are three essential factors often overlooked in current education, and certainly not necessarily inspired by the current digital lifestyle. The first is to have a strong sense of one's own individual identity (and to respect it in others). The second is to have a sense of individual fulfillment. The third is to be useful to society. How might these somewhat abstract goals be realized?

There is something that ticks all three boxes: creativity. By creativity I don't necessarily mean writing a symphony or revealing some great new insight into science or the human condition, although such activities would of course qualify. On a more basic level, surely the essence of creativity is simply seeing something in a new way, whether it be rearranging the bedroom furniture or interpreting a social situation from a different angle. Let's unpack the idea further.

Creativity is often associated in particular with young children. It is also associated by some, such as the clinical psychologist Louis Sass at Rutgers,[23] with schizophrenia, and by others (usually the individuals themselves) with the taking of recreational drugs. But not all children, schizophrenics, or drug takers are overtly creative, nor do creative people have to be young, mentally ill, or doped. The clue here might lie in the fact that some of the features exhibited by children, schizophrenics, and drug takers could be a necessary but not sufficient requisite for creativity. Meanwhile, the same condition of creativity may well be attained by people who fall into none of these three categories. What could this first requirement be?

Young children, as we saw, have sparse brain connectivity, so they cannot readily see one thing in terms of anything else. Schizophrenics resemble children in taking the world literally and not being able to interpret proverbs; in both cases what they get is what they see. Finally, as a consequence of the psychoactive substances impairing their neuronal connectivity, drug takers have impaired associative powers. For them, meaning is fragile and idiosyncratic. So could it be that the crucial first step in the creative process—but

only a first step—is the ability to dissociate previously convention-
ally connected elements? This kind of deconstruction is familiar in
art, where the whole trick is to reduce a cognitive take on an image,
such as a vase of flowers, to an abstract sensory conglomeration of
colors, shapes, and textures that you then try to reproduce. Similarly in science, the essential first step is to challenge dogma, as, for
example, Barry Marshall did with the notion that ulcers were
caused not by stress but by a bacterium.

However, it's important not just to challenge dogma but also to
replace it with an alternative, a *new* association that has never been
tried before: words combined in a special way, a certain conver-
gence of colors and shapes, a familiar object or person in an unexpected context, or a link between two previously unrelated features
of the physical world, such as the parallels between the immune
system and the idea of Darwinian survival of the fittest, which was
first pointed out by the brilliant Australian immunologist Frank
Burnet.[24]

But such a process of deconstruction and reconstruction does
not guarantee a creative act, as anyone experimenting with odd ingredients in a new culinary concoction will testify. Another example would be a child's painting, where there may well be unusual
colors or shapes depicting an animal or a person but the final work
wouldn't qualify for exhibition in an art gallery. The crucial final
step toward creativity, as I see it, is that the work or idea should
mean something, that it should help you see the world in a new
way. Whether through science, art, literature, or any other medium, new connections are established in the brain that in turn
give the world a new meaning. For connections to have meaning,
as we have seen, they cannot be just random: they need to link to
ever wider conceptual frameworks that give a correspondingly ever
deeper meaning.

Creative thinking cannot be purchased, downloaded, or guaranteed, but it can be fostered with the right environment. Developing individual conceptual frameworks for understanding and

interpreting the world also means encouraging individuals to have the confidence to question and deconstruct dogma and traditional views, to possess the courage to make new associations without fear of the opinions or cynicism of others. It is not a happy scenario to imagine a world peopled by individuals who have brilliant sensorimotor coordination, can multitask well, and get high scores on IQ tests but are incapable of reflective thought and understanding, let alone original ideas.

In 1964, at the New York World's Fair, the science fiction writer Isaac Asimov came up with this appropriate, enormously prescient prediction for fifty years hence:

> Even so, mankind will suffer badly from the disease of boredom, a disease spreading more widely each year and growing in intensity. This will have serious mental, emotional and sociological consequences, and I dare say that psychiatry will be far and away the most important medical specialty in 2014. The lucky few who can be involved in creative work of any sort will be the true elite of mankind, for they alone will do more than serve a machine.[25]

Then again, perhaps to people in the future, the priorities of the ancient thinkers, of visionaries such as Asimov, and certainly of typical Digital Immigrants like myself, will seem as obsolete, as risible, and as inappropriate for the mid-twenty-first-century agenda as the mindset of the Victorians was to that of the twentieth century. Still, we cannot ignore the real world. However much digital technologies draw us into their pixelated, frenetic hall of mirrors, this world still serves as a parallel to the ever present, bulky, three-dimensional environment in which even the geekiest technophiles still have to exist.

MIND CHANGE
BEYOND THE SCREEN

I n Shakespeare's time, someone forty years of age was considered old. In jaw-dropping contrast, a baby born today has a one in three chance of living to be a hundred years old, at least in our privileged developed world.[1] Diseases such as polio and diphtheria are now specters of the past, with new advances in medicine raising our expectations of good health ever higher. Meanwhile, whole new branches of medicine, such as gene therapy[2] and regenerative medicine,[3] are opening up wonderful possibilities.

How will the existence of these pioneering medical technologies impact the twenty-first-century mindset? Upcoming generations will probably take these advances for granted, just as we baby boomers never regarded polio or TB as serious health threats in the way that our own parents did. And further back, in the early decades of the twentieth century, most people would have accepted discomfort at best and pain at worst as the norm, whether from a rotting tooth, cataracts, joint pain, or infection. Nagging minor ailments would have been a way of life, and the brain would have adapted, as is its evolutionary mandate, to whatever ongoing situation presented itself. But then again, if it was the default to be physically uncomfortable, people would not have been able to reflect so readily on themselves and their lives in the way that is possible today. Moreover, the highly plausible likelihood of some capricious and indiscriminate illness suddenly wrecking your life, or that of someone close to you, would have overshadowed your daily existence. Nowadays, such fears are receding, and in the future biomedical technologies might further encourage the belief that good health is the birthright of the human species.

However, there is one disease, or rather range of diseases encompassing one particular dreaded symptom, that is more devastating than any other. If we are concerned about Mind Change, then we also need to think about mind loss—not just through mindless screen activities, but more permanently through brain disease or dementia, literally a "loss of mind." As we've already seen, if the mind can be regarded as the personalization of neuronal connections, then the gradual dismantling of those connections would be the physical process that underlies the confusion and loss of memory that characterizes diseases such as Alzheimer's. By the middle of the twenty-first century, two million people in the United Kingdom alone will be suffering from Alzheimer's disease, which accounts for about 70 percent of the instances of dementia.[4] Think about how many people love you in the world. For the ease of the ensuing math, let's say ten; that means there will be twenty million lives turned upside down, about a third of the British population. It was estimated that in 2010, 35.6 million people were living with dementia worldwide, with numbers expected to almost double every twenty years, to 65.7 million in 2030 and 115.4 million in 2050.[5] In a 2013 U.S. study, dementia emerged as a more expensive economic burden to society than heart disease or cancer.[6]

Dementia is a singularly cruel affliction in its devastation of such a large number of lives. Although heart disease, say, or cancer can be life-threatening diseases, the patient is still the same person he or she always was, still aware of being your husband or wife, mother or father, brother or sister, and therefore still having a meaningful relationship with you despite the illness. Not so with dementia.[7] As the disease takes its remorseless toll with the slow yet continuous loss of brain cells, so the caregiver can undergo indescribable distress as an afflicted parent or spouse may deny any relationship. The sense of loss can be every bit as sharp as if their loved one had actually died or been killed. Caregivers often undergo all the signs and stages of bereavement, but without receiving the consideration and allowances society normally affords those suffering a personal loss.

There is currently no effective treatment for the spectrum of neurodegenerative disorders characterized by dementia.[8] But let's assume, and indeed hope, that sooner or later someone can come up with a breakthrough. Imagine going to your primary care physician for a routine blood test, just as you might to check your cholesterol levels, and the doctor calmly looks you straight in the eye and says, "Well, there's good news and bad news. The bad news is that you have an elevated biomarker for neurodegeneration in your blood. This means, according to the chart here, that in your case, in about two years, certain symptoms will appear: short-term memory difficulties or problems finding the right word for an everyday object. However, the good news is that we now have an oral medication that will stop any more of your brain cells dying. So start taking these tablets today. You'll need to take them every day from now on, but as long as you do, you'll never experience any symptoms, because we'll have stopped the neurodegenerative process in its tracks." This scenario of a routine blood test and daily medication could eventually be a serious reality rather than a fantasy. The crucial bit of knowledge still needed is what makes specific cells in the brain embark on the cycle of cell death that we call neurodegeneration.[9] The identification of this basic mechanism underlying Alzheimer's and related diseases is the holy grail that will then lead to an early (ideally presymptomatic) diagnosis and the all-important medication for preventing any more cells from dying.

So let's assume that this wonderful prospect is realized and that dementia eventually joins those other diseases of the past that once were, or seemed to be, death sentences but are now containable thanks to new biomedical strategies. By the second half of this century, many of us will be looking forward to a long and healthy life. We will also look younger as a result of being healthier, and we will be able to reproduce for much longer, perhaps eventually for our entire lives. As the technology improves, it could even become the norm for a woman to have her eggs frozen when she is in her reproductive prime, to be thawed later, perhaps even when she is post-menopausal, so that she can have a child, albeit by in vitro fertilization. Let's take the scenario further to an extreme. Unpalatable and far-fetched though it might sound, it is not beyond reasonable scientific expectation that in the future anyone, regardless of gender, age, or sexual orientation, could have a child with anyone else. If it becomes possible to extract the genetic material from any cell in the body and combine half of it with someone else's, there will be no further need of sperm or egg.[10] What would be required would be an evacuated egg and a womb, which could be supplied by different people. Therefore, in principle, a child could eventually have six parents: the genetic donors, the donor of the egg, the donor of the womb, and the two parents who raise the child. The main point is that, one way or another, you could be a new parent, caring for a small baby, throughout adulthood.

Finally, there's work. Traditionally, paid work was outside of the home and often entailed physical fitness. Now that the knowledge economy and the cyberworld have made working from home possible and physical strength and mobility no longer essential, there is a growing argument against having a fixed retirement age; indeed, this is now becoming the case in many organizations and societies. If we weight the case further with the idea that stimulation of the brain is much better for you than passive disengagement from the outside world, then work might even be sold to society as being good for the brain. Suffice it to say that if we have an increasingly aging sector that is mentally agile and healthy, retiring on a pension to play golf or Sudoku will be the exception rather than the norm.[11]

When you first meet someone, you probably, if subconsciously, allocate them to a particular generation on the grounds of (1) how healthy they seem, which will impact on (2) what they look like, (3) their reproductive status, and (4) whether or not they are still working. If the biotechnology-driven trends now in motion play out to the logical conclusions outlined above in all these four crucial areas of our lives, such compartmentalization into one generation or another will not be so easy.

So much for a changing outside world—but one grounded in a good old-fashioned 3-D physical reality as opposed to the 2-D cyberlife. Yet now imagine if the two were to merge. What if the digital technologies previously confined to the screen could affect the way you experience the real world? Nobody nowadays wants to remain tied to a clunky keyboard and a cumbersome separate screen. Already, smartphones are handheld computers that happen to have a phone facility; and there is a burgeoning preference for mobile devices rather than laptops that offer all manner of apps and videogames. The next generation of smartphones will be context-aware, exploiting the growing availability of embedded physical sensors and data exchange abilities. As a result, phones will start keeping track of your personal data, and will adapt to anticipate the information you need based on your intentions and location.[12] As well as monitoring you, the phone will monitor your surroundings and reveal information about anything or any place at which you point your phone.

Now imagine what life might be like if the much valued, protective feature of texting could be oral rather than written. You'd avoid the difficulty and embarrassment of interacting directly with someone, even on the phone, by recording messages which are then accessed as swiftly and easily as written text messages are today. You wouldn't even have to be literate. This new invention would create a firewall between you and the squalor of real, immediate human contact, along with the growing disaffection with laborious reading and writing skills. Welcome to the world of Google Glass.[13]

In appearance, Google Glass looks just like a pair of normal spectacles with a small black oblong at the top on one side, which shows you information and through which you can access the Internet via simple voice commands. Soon you will be able to record whatever you see, hands-free, and share it in real time with others. Moreover, you'll be able to get directions and find out anything about your current location, have your words translated, and receive information about wherever you are, without even asking explicitly.

Until now, sessions of gaming, social networking, and Web surfing all come to an end at some point. You can always turn the device off and walk back into the real world. Google Glass and other such technologies will make most of these activities possible every waking moment. Just as it's now commonplace to see passersby with wires snaking into their ears talking loudly to themselves—people who would once have seemed plain loony—there's now the prospect that these same people will have morphed into a species with minimally notable rimless specs, living in the Google Glass "augmented reality."[14] The technology functions by *enhancing* one's current perception of reality and therefore should not be confused with virtual reality, which replaces the real world with a simulated one. Instead, augmented reality is an ongoing view of a physical real-world environment whose elements are "augmented" by computer-generated sensory input such as sound, video, graphics, or GPS data.[15] In this way, "artificial" information about the environment and what it contains will be constantly overlaid on the real world.

The plan is for Google Glass to be launched in 2014, so you might even be reading these words right now with your own Google Glass ready at hand, longing to put it back on. In any event, the implications of mass adoption of this new way of seeing the world are as diverse as they are profound. Predictions for wearable computing devices such as Google Glass or Apple's proposed iWatch are that they will become the norm for most of us within five years, with 485 million annual device shipments by 2018.[16]

Once you're all wired up to an augmented reality, just imagine how terrible it will be to be on your own, how hard it would be to abandon this new dimension and just switch everything off. Already, the majority of phone owners are emotionally attached to their smartphones. In one 2012 study of U.S. phone users, 73 percent said they felt "panicked" when they misplaced their phones; 14 percent said they felt "desperate," while 7 percent felt "sick."[17] In the United Kingdom, 66 percent of phone users reported a fear of losing their phones, which now even has a name, "nomophobia."[18] If this type of attitude already exists, then it's breathtaking to contemplate the type of emotional attachment we might have to intensively integrated devices that provide more entertainment, faster answers, and even more sanitized socialization, all seamlessly.

It is hard to understand how the human brain will absorb such a tsunami of information. With Google Glass, you will have the facts in your face without the need to try to work out the answer yourself. If search engines are already offering a faster and easier option than taking your brain through its otherwise necessary workout, you'll now run the risk of becoming mentally flabby in a way that isn't even possible at the moment, as surfing on a mobile phone or tablet or laptop still requires some proactive typing or touching. You will no longer be driving what you look at: the display will be driving you. The most immediate feature of Google Glass is its interactivity, with the emphasis on the constantly updated present moment. This constantly ongoing literal world will permanently trap users in an endless hyperconnected present. There will be nothing private to remember or anticipate.

Google Glass could also sound the final death knell, once and for all, for privacy. Andrew Keen, who describes himself as "a

British-American entrepreneur, professional skeptic and the author of *The Cult of the Amateur* and *Digital Vertigo*," has been quick to flag this issue. "These glasses, a kind of digital surrogate for our eyes, are strange in a creepy, Hitchcockian, *Rear Window* sort of way," he writes. "Or the same way that Big Brother's ubiquitous cameras were strange in George Orwell's *Nineteen Eighty-Four*. And in the same way that a future in which 'promethean' data companies like Google rule the world now appears strange." He continues:

> But Google Glass opens an entirely new front in the digital war against privacy. These spectacles, which have been specifically designed to record everything we see, represent a developmental leap in the history of data that is comparable to moving from the bicycle to the automobile. It is the sort of radical transformation that may actually end up completely destroying our individual privacy in the digital 21st century. When we put on these surveillance devices, we all become spies, or scrooglers, of everything and everyone around us.[19]

So, here is a truly new type of future straight out of science fiction, where the currently nascent obsessions with monitoring the lives of others and broadcasting every moment of your own existence are now finally liberated completely from keyboard and touch screen. Instead you're truly interfaced directly with a digital device: it is an extension of your body. My concern is not only with the ethics and legality of the possible loss of privacy, as it is for Keen, but also with what such a loss will mean to us as the independent individual entities we've all been until now.

Google Glass wearers may well feel pressured to opt into the hyperconnected cyberworld all the time for fear of otherwise missing out or of being left behind. The trade-off for the resulting disclosure in what they are doing every minute of the day is, as it always has been, loss of privacy. Until now, privacy has been precious because it has been the other side of the coin to our identity. We see ourselves as individual entities, in contact with the outside world for sure, but at the same time always distinct from it. A sense of privacy keeps the two apart. We don't disclose certain facts about ourselves, not because we're ashamed or embarrassed by them, but quite simply because we feel that not everyone should know what we are feeling or thinking. *By holding back, we are preserving a sense of self distinct from the outside.* Privacy provides the boundary: it stops us being transparent. This is why most of us draw the curtains at night to prevent strangers from looking into our homes. We interact with the outside world, yes, but always in dialogue with our brain, something to which only we are privy. Thus we always have an inner narrative, an ongoing thought process that is ours alone, a *secret* life—until now.

If you're now trapped in the present, constantly catering to the demands of the outside world, that inner narrative might be harder to sustain. Your secret sense of identity might become less and less important, less meaningful, because it no longer has the all-important context, the inner conceptual framework where one event, object, or person relates to another according to your own unique framework of connectivity. The *you* now externally constructed by Google Glass may not allow much time and opportunity for those internal memories, the secret reflections, to develop and blossom fully. But if privacy were needed only to protect this inner awareness, if there were no longer a secret life anyway, then privacy is meaningless. In contrast, if you define yourself by the degree of attention you receive from others, the loss of privacy is to be welcomed in order to permit a completely new type of identity: a connected one.

Let's go one step further. What if you were integrated with the outside world all the time? Perhaps this would lead to a kind of life where the secondhand thrill of self-reporting, posting feedback, and receiving feedback completely trumps the experience itself. Your identity is now paradoxically *online* moment to moment but essentially *offline* in that its importance lies in its reporting. The excitement you feel is generated not by the raw firsthand experience but by the slightly delayed, indirect, and continuous reaction of others.

If we live in a world where face-to-face interaction becomes uncomfortable and where personal identity is increasingly defined by the approbation of a virtual audience, the most personal relationships might change as well. It will be a hard transition for an individual accustomed to an audience of five hundred "friends" who share a collective flood of consciousness to switch to a one-on-one long-term relationship that is exclusive and completely private. Interestingly, two of the most technologically advanced collective-minded countries today (Japan and South Korea) are facing huge problems in declining birth rates.[20] Of course, any decline in interpersonal skills for conducting deep and meaningful partnerships doesn't necessarily imply a comparable decline in sex; it may be a relatively straightforward process to extrapolate from the sensory-laden sexual adventures of a videogame to similar experiences in real life. Sex would now be more casual, less meaningful, and highly transient.[21] On the other hand, perhaps even sex itself, involving as it does even at the level of the basic act, issues of self-confidence, trust, and vulnerability, may also develop into an aversion. Once again, evidence from Japan and Korea indicates a lack of interest in sex, or even in dating, among the younger generations. Tellingly, nearly half of all Japanese women ages sixteen to twenty-four are "not interested in or despise sexual contact," and nearly a quarter of men feel the same way.[22]

Another ramification of the technologies on the move will be the abandonment of the sedentary cyberlifestyle. A trend we are already seeing is that a screen life may be leading to an understimulation not only of the sense of touch but also of taste and smell, which in turn could be a factor leading to ever more indulgence in eating and drinking. The prevalence of obesity in England has more than tripled in the last twenty-five years. The latest Health Survey for England data show that in England in 2010, 62.8 percent of adults (age sixteen or over) were overweight or obese and 30.3 percent of children (ages two to fifteen) were overweight, with 26.1 percent of all adults and 16 percent of all children crossing the line into obesity.[23] While there are many complex reasons for this alarming rise, including a poor diet of cheap junk food high in sugar and calories, another definite factor is insufficient physical exercise, which can be linked to a life spent sitting in front of a screen. Surely here the mobile technologies will offer at least the obvious advantage of a reduction in obesity through increased movement. But an alternative perspective is that the drive for stimulation of the senses of touch, taste, and smell, to which the screen does not cater and which may be too risky to achieve in a close physical relationship, may be met by further eating, which would offset any potential reduction in obesity.

So there you are, weaving among the crowds but oblivious to them. At least one of your hands is holding something easily edible, and in your ear there's an incessant stream, perhaps of music but more likely previously recorded oral text messages, or perhaps information as to where you can buy the latest goods that your personal traffic history has revealed are just right for you. Cyberspace is no longer limited to the two-dimensional screen but extends to three dimensions, thereby transforming reality. Your world is more like a bubble. Outside other people are passing by, but you are protected from them by the transparent shield encompassing your virtual space, your new dimension. You can touch and smell things, as well as hear and see them, but you are never alone, never independent. Always there's the voice in your ear, your best friend, acting as intermediary and therefore paradoxically distancing you from everyone and everything else at the same time as it connects you.

Bear in mind that you will have no strong sense of who you are, no sense of past or future, just the atomized moment. You'll be in a continuous state of high arousal, craving novelty and stimulation as each input is evaluated on purely (literally) sensational terms and thus soon palls. You will be very vulnerable to manipulation, both in how you see the world and in how you react to it. Like a small child, you'll readily obey and conform, since you have adapted to expect and prioritize the constant approval of others. So you'll be grateful for the voice, since you may be a little confused. After all, you'll no longer have a conceptual framework for understanding what is happening around you. Added to this blurring of self and outside world will be a blurring of fact and fantasy. Since you are no longer just using your senses, but everything is aided and abetted by your cyber best friend, the boundary of reality will be increasingly smudged, as is your now ambivalent generational status, thanks to advances in biotech. The three age-old distinctions that formed the basic constructs of our lives—private inner self versus external others, fact versus fantasy, and child versus parent versus grandparent—may for the first time start to erode.

An extreme and far-fetched scenario? Of course. Yet none of these future developments is a sci-fi fantasy on par with time travel, say, or perpetual motion machines. *They are all starting to happen right now.* These and similar technologies will have enormous and far-reaching implications for how the next generations will behave and, most important, how they will think during their long and healthy lives. As I have said, the critical issue facing us is one of transition. How do we not only make sense of but flourish in the current question-poor, answer-rich technological blizzard? For those born in the second half of the twentieth century, extraordinary advances in resources, health, and culture have increased our life expectancy compared to those born a generation or so earlier. As a result, larger swathes of the developed world have more options, more privileges, and more time in which to explore their full potential. So how can we ensure a future where our technology does not frustrate, but actively fosters, deep thinking, creativity, and real fulfillment?

20

MAKING CONNECTIONS

Just think back to a decade ago, when there was neither Facebook nor Twitter, and when Wikipedia had fewer than fifty thousand articles instead of the excess of four and a half million available today. And could anyone have predicted in the early 1980s that within just a few decades, six billion of the seven billion people in the world would have access to a mobile phone, while only four and a half billion have access to a working toilet?[1] The past was indeed a foreign country, as L. P. Hartley observed: they did things differently there. So what will the new country of the mid-twenty-first century look like? More significant, how would we like it to look?

Some sneer at any attempt to predict the future. The seeming arrogance of previous generations can appear ridiculous and naive with the glory of hindsight. One often quoted and apocryphal example is that of Thomas J. Watson, former head of IBM, who foretold that, at best, there might be a market for five computers in the world.[2] While this shows that predictions about the long-term consequences of particular inventions are uncertain, carrying the basic scientific *concepts* a step further can raise interesting questions about the world we are creating. While we may not be able to predict consumer enthusiasm, we can contemplate where new technologies could lead if taken to the extreme. For example, George Orwell's *1984* envisaged a world of surveillance and manipulation of thought, where an omnipresent Big Brother ruled absolutely. This book remains a classic because it suggests eerie parallels to our world today.

"[Might man become] a mere parasite of machinery, an appendage of the reproductive system of huge and complicated engines which will successively usurp his activities?" You might think that this is a quote from Richard Watson, Nicholas Carr, Larry Rosen, or another of the current thinkers cited in these pages. In fact, the quote is from a paper in 1923 by the brilliant biologist J. P. Haldane, delivered to the Heretics Society at Cambridge University.[3] Haldane entitled his paper "Daedalus" after the father of Icarus in Greek mythology, referencing him as the creator of the Labyrinth, home to the Minotaur. Haldane's purpose was to focus on the terrible consequences of our own cleverness.

Still reeling from the horrors of the mechanized butchery of the First World War, he explored the future of science, which he described as "the free activity of man's divine faculties of reason and imagination." Many of the predictions in "Daedalus" not only are spookily prescient but also articulate fears that resonate with the worries we've explored in the previous chapters. Although he envisaged that chemistry would continue to transform life with explosives, dyes, and drugs, it was in the application of biology that Haldane foresaw the big transformations. Looking hard at the nascent eugenics movement of the time, he wondered if this might result in "eugenics officials" and "marriage by numbers."[4] These are prospects that may come to pass with the advent of genetic screening and online dating, respectively.

Along with foreseeing our ability to cure many infectious diseases, Haldane predicted the development of a "nitrogen-fixing" plant, which anticipated genetically modified food. Perhaps even more impressive is that he actually prophesied the development of IVF and the complete dissociation of sex and reproduction. So troubling and fascinating were these ideas that they inspired Aldous Huxley to write his famous dystopian novel *Brave New World,* which anticipated a Central London Hatchery for babies and the worst consequences of genetic manipulation. Haldane was also uncannily on target in predicting hormone replacement therapy: "This change seems to be due to a sudden failure of a definite chemical substance produced by the ovary. When we can isolate and synthesise this body we shall be able to prolong a woman's youth, and allow her to age as gradually as the average man." Even mood-bending drugs were within his vision, "to control our passions by some more direct method than fasting or flagellation." This idea was also appropriated by Huxley: the citizens of his dystopia routinely take "soma" pills and become immediately and unconditionally ecstatic.

Haldane listed in "Daedalus" the big questions of science that are still with us today: "first of space and time" (in our terminology, the Big Bang), "then of matter as such" (for us, the persistent quirkiness of quantum theory and the dream of nanoscience), "then of [man's] own body and those of other living beings" (surely the synthesis of different branches of biomedical science), along with the great question of how a brain can generate the subjective experience of consciousness, "and finally the subjugation of the dark and evil elements in [mankind's] own soul." At a stretch, the biggest question is of how we shall use this knowledge to work out the degree of biological determinism, the ultimate question of free will in the digital age.

Granted, we may not be able to predict the precise technologies and consumer products of the future. But it *is* possible, as Haldane, Huxley, and Orwell have shown, to articulate the underlying scientific idea, observe its current manifestation, and predict where such technology might be headed in its possible impact on human existence, society, and mindset. The preceding chapters have given snapshots of where we are at the moment, but the most important question of all is where these new developments could lead us if they continue unabated and unfocused. Sleepwalking into the unknown, proudly unprepared and just hoping for the best, is surely the most perilous option. By letting our imaginations unfold and by looking toward more distant horizons, we admittedly run the risk of straying into mere speculation: but proactive thinking allows us to take critical stock of our world today and puts us in the best possible position to devise a game plan for an optimal future.

Humanity has always had a love-hate relationship with "progress," in equal measure delighted by the convenience a new invention brings and worried that it might just rob us of some quintessential quality. Some four hundred years before the birth of Christ, Socrates was concerned that writing would destroy mental prowess, with arguments eerily similar to those we've explored here with respect to the Internet. He argued that writing

> will create forgetfulness in the learners' souls, because they will not use their memories; they will trust to the external written characters and not remember of themselves. The specific which you have discovered is an aid not to memory, but to reminiscence, and you give your disciples not truth, but only the semblance of truth; they will be hearers of many things and will have learned nothing; they will appear to be omniscient and will generally know nothing; they will be tiresome company, having the show of wisdom without the reality.[5]

These perennial worries were dramatically inflamed at the beginning of the twentieth century when the mass adoption of automation became a recognizable force in our lives, along with the electricity that powered it. The underlying plot is perhaps obvious and readily seen as Romantic: despite the benefits it brings, mechanization will somehow rob us of all the less tangible but most basic features that we hold dear about our species, namely, our emotions.

An early illustration of suspicion of the pitiless robot is captured in the 1927 German expressionistic film by Fritz Lang, *Metropolis,* which plays to the fear that technology will dehumanize us. In the film (a visual masterpiece combining art deco and industrial imagery) we see the horrors of mechanization through the eyes of Freder, the spoiled son of the owner of the large industrial city, as he discovers how the workers are effectively treated as machines.

Another dim vision of the future is evoked in Edwin Muir's 1952 poem "The Horses," which describes the immediate aftermath of a world destroyed by technology and how the survivors embrace the old traditional way of life. The speakers refuse to return to

> *That old bad world that swallowed its children quick*
> *At one great gulp . . .*
> *The tractors lie about our fields; at evening*
> *They look like dank sea-monsters couched and waiting.*
> *We leave them where they are and let them rust:*
> *"They'll molder away and be like other loam."*[6]

A third fictional scenario in which technology poses a threat to humanity, and perhaps the most familiar, is that of the robot HAL in Stanley Kubrick's 1968 classic, *2001: A Space Odyssey.* HAL is capable of speech, speech recognition, facial recognition, natural language processing, lip reading, art appreciation, reasoning, chess playing, and, most unsettling of all, interpreting and reproducing emotions. When eventually the astronaut Dave begins to remove his modules one by one, HAL's consciousness slowly disintegrates in a way that seems painfully human, going all the way back to childhood songs:

> I'm afraid. I'm afraid, Dave. Dave, my mind is going. I can
> feel it. I can feel it. My mind is going. There is no question

about it. I can feel it. I can feel it. I can feel it. I'm a . . . fraid. Good afternoon, gentlemen. I am a HAL 9000 computer. I became operational at the H.A.L. plant in Urbana, Illinois, on the twelfth of January 1992. My instructor was Mr. Langley, and he taught me to sing a song. If you'd like to hear it I can sing it for you.

In real life the reverse scenario can also send a chill up our spines: not so much machines trying and failing to be human, but humans trying to escape the ravages of emotion by trying to be machines. In 1959 the child psychologist Bruno Bettelheim published "Joey: The Mechanical Boy," a case history of a very disturbed little boy who converted himself into a robotic-type entity as a defense against the world. At the end of the article, after successful treatment, Joey makes a banner for the Memorial Day parade with the words "Feelings are more important than anything under the sun." "With this knowledge," concludes Bettelheim, "Joey entered the human condition."[7]

But could this now be the very way we are headed—into a mechanized future devoid of all the human qualities we cherish? If enthusiasm for technology has usually been tempered in previous generations with worries that it could be dehumanizing, it seems that Digital Natives typically do not have such qualms. To summarize the preceding chapters: social networking sites could worsen communication skills and reduce interpersonal empathy; personal identities might be constructed externally and refined to perfection with the approbation of an audience as priority, an approach more suggestive of performance art than of robust personal growth; obsessive gaming could lead to greater recklessness, a shorter attention span, and an increasingly aggressive disposition; heavy reliance on search engines and a preference for surfing rather than researching could result in agile mental processing at the expense of deep knowledge and understanding.

These snapshots may seem a bit unfair in that their brevity boils down complex differences across cultures, generations, and individuals to simplistic caricatures, but they can be useful for us to reflect on where they might lead. Interestingly enough, the profile emerging from this list is *not* of a ruthless robot but of an all too human mindset amplified in all its frailty and vulnerability, craving attention as a unique individual and at the same time, paradoxically, needing desperately to belong and to be embraced within a collective identity and mindset. When feelings and emotions are amplified and constantly held at a premium, it's no wonder Digital Natives don't have the age-old concerns about mechanization robbing them of their humanity. So what exactly is problematic about a culture that taps into such deep-seated biological needs?

The difficulty arises when, thanks to the unprecedented nature of the digital lifestyle, these natural tendencies become exaggerated as never before. The first basic human need is to be acknowledged as special. We've seen how narcissism is increasing thanks to social networking sites. Indeed, only recently a word was reported as increasing in frequency 17,000 percent since it was first used in 2002: "selfie," the taking of your own picture.[8] Without the handbrake of body language that usually constrains interpersonal communication, and with access to a community larger than any group of real-life friends could ever be, the drive to be someone special could get out of hand, even become obsessive.

The cyberlife obligingly offers an unprecedented means of achieving status, measured for the first time ever not by possessions, talent, or job. Without question, the excessive flaunting of such culturally acknowledged traditional signs of status can be pernicious and detrimental, as argued persuasively by Oliver James in *Affluenza,*[9] but status is now simply measured by how "cool" one is, how many followers and friends one can attract in cyberspace, and not by one's abilities or achievements. Add to the mix the unprecedented opportunity for concealing the real self, and the possibilities for an individual never to feel at ease in meaningful face-to-face relationships are even greater. The answer is to retreat instead into the safe world of the screen in the quest for approval, having done little to earn it, and indeed not even existing in the same way that people do in the real world.

The second natural human desire is to be accepted as one of the tribe, to be a small part of a larger collective identity. Once again, the screen world can cater to this need on an unprecedented scale in that you can join with others without the real-world effort or skills normally required to be in a choir or on a soccer team and even without the physical effort of going to a soccer match as a supporter. But while soccer teams and choirs generate an objective end product, and even purely fun hen or stag nights have a specific ending, an Internet community has the boundless time, numbers, and lack of accountability to turn in on itself and develop a collective identity that may or may not be for the good. Bertrand Russell remarked in his reply to Haldane, "Icarus," that "men's collective passions are mainly evil,"[10] while behavior on 4chan has been likened by one of the site's devotees to that of the schoolboys turned savages in William Golding's *Lord of the Flies.* A recent example of a collectively negative mindset online is the case of the woman threatened with rape on Twitter, not just by one deranged man but by a whole following. Her crime? Suggesting that since no woman is currently featured on any British banknote, the universally acclaimed author Jane Austen should grace the latest edition of the £10 note. Once the hue and cry was raised, a mob mentality took over.[11]

A final and third aspect of being human, which is nonetheless exaggerated in cyberspace, is our impulsivity, the desire for instant gratification. We've seen that a key attraction of videogames is that actions do not have long-term consequences, but this is just the tip of the hedonistic iceberg. The sheer pleasure not just of playing games but of watching YouTube or of disclosing everything about yourself on Facebook trumps any long-term implications. Pandering to this, indeed providing an excuse at the end of any account of reckless and often thoughtless actions, is the simple and widely used term "YOLO" (you only live once), appended to a description, for example, of outrageous or excessive behavior as justification or explanation.[12] By focusing on the moment, by being the passive recipient of a sensational time, you've "let yourself go." The big difference between now and the recreational abandonment of previous generations is that now you can do it much more often, on demand—almost all the time if you want to.

It just might be that the cyberculture enables you to satisfy all three basic drives more fully and more easily *in combination* than at any other time in human history.[13] Think back to the imagined scenario in Chapter 19 in relation to mobile technologies. First, there is the strong sensory stimulation of the exciting audiovisual inputs that distracts you from thinking ahead or reflecting on the past (YOLO). Second, at the same time, you're going to be connected, and increasingly hyperconnected, say by Google Glass, and therefore will always be one of the tribe. Third, you are constantly acting and reacting as a single player before your audience: you require constant feedback from them, paradoxically living a life indirectly, at secondhand and offline, but in an online mode of constant readout that, if you're "cool," brings you recognized status as someone special.

So instead of the digital age being just like earlier technologies in posing the age-old threat of dehumanizing us, of pandering to the perennial fear that scientific and technological progress will turn us all into zombie-like cyborgs, I suggest that the *exact opposite* is the case. Some of the very worst aspects of being all too human—the desire for status irrespective of talent, mob mentality, and uncaring recklessness—are now being given free rein throughout the uncharted territory of cyberspace. What can or should we do?

In Douglas Adams's *Hitchhiker's Guide to the Galaxy,* a group of hyperintelligent pan-dimensional beings demand to learn the "Answer to the Ultimate Question of Life, the Universe, and Everything" from the supercomputer Deep Thought. Deep Thought then needs seven and a half million years to compute and verify the answer, which turns out to be 42. But note that even here the Ultimate Question itself is unspecified. Even though many have articulated comparably ambitious thoughts on the meaning of life through the centuries, they still amount to a tiny minority privileged with the leisure to continue to contemplate the significance of who they were and what they were doing, while still fewer have had the opportunity to express their reflection in literature, music, art, or science. But we are now entering a time, potentially, of real opportunity to stretch ourselves en masse where each one of us realizes their individual, true potential, to ask big questions and to develop original and exciting solutions.

Before we get too carried away with the prospect of such a rosy future, we need first to decide what our priorities are, what kind of society we want, and what kind of individual traits we value. To this end, the traditional print and broadcast media could get the ball rolling. After all, they can access the broadest range of different types of people, not just the overheated blogosphere; moreover, unlike those ranting in cyberspace, they have the legal obligation to be accurate. Debates and in-depth interviews with a range of specialists would ensure that everyone has access to as many views and insights as possible. Perhaps someone might even think of making a film. After all it was *An Inconvenient Truth* that woke up the silent majority of us to climate change.

Step two would be to take the pulse of societies around the world. It would be really helpful to have formal surveys of the views of stakeholders such as parents and teachers, psychiatrists and neuroscientists, as well as the Digital Natives themselves. As we saw back in Chapter 1, the kinds of surveys published so far have been mainly simple numbers, statistics, and demographics. Now we need surveys that go beyond raw statistics to canvass the opinions of all sectors of society. Mind Change encompasses questions that could be every bit as complex and varied as those around climate change. But a big difference is that while most people would prefer the planet not to overheat, the possible and desired outcomes from

Mind Change may vary widely according to different tastes and predilections, so we need to look at the full spectrum of views out there.

The next key issue is the thorny one of evidence. While specific experiments in the lab can and have been designed to answer specific questions, further funds and resources are still needed from both the private and the public sectors for more basic lab-based research, epidemiological studies, and psychological and sociological approaches; this constitutes step three. As we've seen in the previous chapters, the scientific method can be tricky, with more questions being raised than answered by the findings. Much more clarification and detail are needed, as well as simply more data. It is only by investigating on all levels, from the molecular to the societal, how the human brain is developing over months and years that we can assess the real long-term impact of the new technologies on how an individual thinks and feels. The longer we wait before commissioning this type of work, the fewer the options and the narrower the scope we may have in the future. We need to start right now.

Further, there is no reason why the technologies discussed here shouldn't be part of the solution. Step four would be the invention of completely novel software that attempts to compensate for and offset any possible deficiencies arising from excessive screen-based existence.[14]

Of course, these four steps are not really steps at all, in that one is not dependent on the other. Rather, these very different strategies could all be deployed simultaneously.

We come back to H. L. Mencken's famous quote: "For every complex problem there is an answer that is clear, simple, and wrong." Never has that been more true than for the complex situation generated by the all-pervasive cyberculture of the early twenty-first century. The snapshots captured in these pages, which, taken together, constitute Mind Change, amount to a phenomenon whose enormous size and impact make it comparable to climate change. Both climate change and Mind Change are in our hands; in both cases, it's up to us to be proactive and to do something.

Yet there's a further, utterly crucial difference. For Mind Change there is no answer as such, because there is no clear question or goal. Unlike the unambiguous agenda set by climate change, Mind Change depends on what each of us wants and where we want to go as individuals. Moreover, while climate change involves at best damage limitation, the same is not true for Mind Change. With Mind Change, we have the opportunity to harness the powerful technologies it encompasses to positive and unprecedented, albeit as yet unspecified, ends. If we do indeed wish, in the late futurologist Jim Martin's words, not to ask what will happen in the future but to proactively shape that future, then we should expect no quick Manichaean answers, no catchphrase, no sound bite, no easy collective doctrine. We cannot predict what wondrous new technologies will appear, nor even the developments and rate of advance of those already in train, such as mobile technologies. But we *can* emulate Haldane, Russell, Huxley, and Orwell in discerning trends in how we humans adapt to that technology and how it transforms the way we see the world.

The theme of connectivity might provide a good ending point for this current journey. We have seen that by connecting neurons in a unique configuration, the physical brain is personalized and shaped into an individual mind. It is these connections, the personal association between specific objects and people, that give those objects and people special significance. Our experiences over time give each and every one of us meaningful episodes that in turn contribute to a linear narrative, a personal story whose very unfolding echoes the thought process itself. But as we become increasingly hyperconnected in cyberspace, might not our global environment begin to reflect and mirror the networking in our individual physical brain? Just as neuronal connectivity allows for the generation and evolving expression of a unique human mind, the hyperconnectivity of cyberspace could become a powerful agent for changing that mind, both for good and for ill. Working out what this connectivity may mean, and what we decide to do about it, is surely the most far reaching and exciting challenge of our time.

ACKNOWLEDGMENTS

I would like to thank, more than anyone, Dr. Olivia Metcalf. As soon as I sent her the first draft of *Mind Change* with the tentative request that she might be able to help me with the references, she committed immediately to be "in this for the long haul." This was just as well, since the volume of reports and papers to sift through turned out to be far greater than either of us had anticipated. It is no exaggeration to say that Olivia has completely raised the game of the book. Through the endless email iterations between Oxford and Melbourne she has been more conscientious, constructively critical, and enthusiastic than I ever could have dreamed. However, the book itself would not even have been started were it not for Will Murphy at Random House New York, who had the original idea and approached me with an invitation. I'm enormously grateful to him, to his colleague Mika Kasuga, and more immediately to Judith Kendra at Random House UK for all the detailed help and advice they have given over the last two years. I'd also like to thank my good friend Professor Clive Coen for proofreading the final draft of the manuscript. Finally, it would be impossible for me to write any book at all without the unflagging support and fantastic friendship of my agent, Caroline Michel. *Mind Change* is dedicated to Professor John Stein, FRCP, for backing me unstintingly both professionally and personally for more than thirty-five years. John is the best example possible of the definition of a mentor: "someone who believes in you more than you believe in yourself."

NOTES

CHAPTER 1. MIND CHANGE: A GLOBAL PHENOMENON

1. *Mind Change* poses and answers questions using empirical, epidemiological, testimonial, and anecdotal evidence. While all of these styles of evidence are included in the book, the latter three are used mostly to develop questions, whereas significant weight is given to empirical research to answer them. It is not claimed that the research collated here is either a systematic or an exhaustive review of the literature. Studies published up to July 2013 were eligible for inclusion. Preference was given to meta-analyses and peer-reviewed journal publications in instances where the research field of the topic was established. Preference was given to higher-ranking journals where applicable. For areas of research that are brand-new, and for which few peer-reviewed publications yet exist, less robust literature such as conference proceedings and technical reports were consulted. It is important to remember that the scientific field lags behind technological advances and that the speed at which the cyberworld changes creates significant challenges for this area of research. Where possible, preference was given to studies that used the most current forms of technology. Throughout *Mind Change,* notes contain additional references and comments on various topics. Readers are strongly encouraged to source the papers discussed here and beyond, as *Mind Change* is designed as a current snapshot of the literature only.

2. An Australian digital publisher, Sound Alliance, recently commissioned a national survey of some two thousand people between the ages of sixteen and thirty: Mahony, M. (April 22, 2013). Sound Alliance reveals results of national youth research project [blog post]. Retrieved from http://thesoundalliance.net/blog/sound-alliance-reveals-results-of-national-youth-research-project). Among the participants, who typically had an under-

graduate degree or at least secondary education to the age of eighteen, 53 percent looked to social media, rather than TV or newspapers for their news, while 93 percent used Facebook daily even though 22 percent of them thought it a "waste of time." Meanwhile, 89 percent of respondents said they had not yet found a passion or purpose in life but were still searching for it. Of course this continuing quest may well apply to most of humanity, but perhaps more telling is that it is "FOMO" (the fear of missing out) and "FONK" (the fear of not knowing) that drives them constantly to check their phones for Facebook, Instagram, Twitter feeds, and new emails and texts. Stig Richards, the creative director at Sound Alliance, summed it up: "They have so much information coming in through aggregation, principally Facebook, that they are working very hard to keep up with the constant flow. . . . So they aren't able to attribute time and energy into specific passions, to the extent that maybe people could before social media was so pervasive. . . . The youth of today are living their lives one mile wide and one inch deep." (Munro, K. [April 20, 2013]. Youth skim surface of life with constant use of social media. Retrieved from http://www.smh.com.au/digital -life/digital-life-news/youth-skim-surface-of-life-with-constant-use-of-social -media-20130419-2i5lr.html).

3. World Economic Forum. (2013). *Global risks report 2013* (8th ed.). Retrieved from http://reports.weforum.org/global-risks-2013.
4. Department for Work and Pensions. (2011). *Differences in life expectancy between those aged 20, 50 and 80 in 2011 and at birth.* Retrieved from http:// statistics.dwp.gov.uk/asd/asd1/adhoc_analysis/2011/diffs_life_expectancy _20_50_80.pdf.
5. World Health Organization. (2008). *WHO report on the global tobacco epidemic, 2008: The MPOWER package.* Retrieved from www.who.int/tobacco/ mpower/mpower_report_full_2008.pdf.
6. Schwartz, M. (August 3, 2008). The trolls among us. Retrieved from http://www .nytimes.com/2008/08/03/magazine/03trolls-t.html?pagewanted=all&_r=0.
7. Nisbett, R. E., and Wilson, T. D. (1977). Telling more than we can know: Verbal reports on mental processes. *Psychological Review* 84, 231–259. Reprinted in D. L. Hamilton (Ed.) (2005). *Social cognition: Key readings.* New York: Psychology.
8. Prensky, M. (2001). Digital natives, Digital Immigrants: Part 1. *On the Horizon* 9, 1–6. doi:10:1108/10748120110424816.
9. Keen, A. (2007). *The cult of the amateur.* London: Nicholas Brealey, pp. xiii–xiv.
10. Selwyn, N. (2009). The Digital Native—myth and reality. *Aslib Proceedings: New Information Perspectives* 61, no. 4, 364–379. doi:10:1108/0001253091 0973776.
11. KidScape. (2011). *Young people's cyber life survey.* Retrieved from http://www .kidscape.org.uk/media/79349/kidscape_cyber_life_survey_results_2011.pdf, p. 1.
12. Kang, C. (December 10, 2013). Infant iPad seats raise concerns about screen time for babies. *Washington Post.* Retrieved from http://www.washingtonpost .com/business/economy/fisher-prices-infant-ipad-seat-raises-concerns-about -baby-screen-time/2013/12/10/6ebba48e-61bb-11e3-94ad-004fefa61ee6_story .html.
13. Grubb, B. (December 16, 2013). iPad holder seat for babies sparks outcry. Retrieved from http://www.nydailynews.com/life-style/baby-seat-ipad-holder -sparks-outcry-article-1:1544673.

14. The full debate on the impact of technology on the mind can be found at http://www.publications.parliament.uk/pa/ld201011/ldhansrd/text/111205 -0002.htm.

15. Rideout, V. J., Foehr, U. G., and Roberts, D. F. (2010). *Generation M2: Media in the lives of 8- to 18-year-olds.* Retrieved from http://kaiserfamilyfoundation .wordpress.com/uncategorized/report/generation-m2-media-in-the-lives-of -8-to-18-year-olds.

16. Teilhard de Chardin, P. (1964). *The future of man.* London: Collins, p. 159.

17. Badoo. (2012). Generation lonely? 39 percent of Americans spend more time socializing online than face-to-face. Retrieved from http://corp.badoo .com/he/entry/press/54.

CHAPTER 2. UNPRECEDENTED TIMES

1. Watson, R. (October 21, 2010). Lecture to the Royal Society of Arts [blog post]. Retrieved from http://toptrends.nowandnext.com/2010/10/21/lecture -to-the-royal-society-of-arts.

2. By mid-2013, 56 percent of U.S. adults owned a smartphone and 34 percent owned a tablet computer (Smith, A. [2013]. *Smartphone ownership: 2013 update.* Retrieved from http://pewinternet.org/Reports/2013/Smartphone -Ownership-2013.aspx). In the same year 51 percent of U.S. households possessed a dedicated gaming console (Entertainment Software Association. [2013]. *The 2013 essential facts about the computer and videogame industry.* Retrieved from www.theesa.com/facts/pdfs/ESA_EF_2013.pdf), while 39 percent of U.S. adults in 2012 reported spending more time socializing online than they do in face-to-face time (Badoo. [2012]. Generation lonely? 39 percent of Americans spend more time socializing online than face-to-face. Retrieved from http://corp.badoo.com/he/entry/press/54). The growth in screen use among youth is comparable. In 2012, 37 percent of all U.S. youths ages twelve to seventeen own smartphones, up from just 23 percent in 2011 (Madden, M., Lenhart, A., Duggan, M., Cortesi, S., and Gasser, U. [2013]. *Teens and technology 2013.* Retrieved from http://www.pewinternet.org/ Reports/2013/Teens-and-Tech.aspx). Twenty-three percent of the same group owned a tablet computer. As of 2013, the average U.S. household with Internet connection now contains 5.7 Internet-connected devices, and they are often being used simultaneously (Internet connected devices surpass half a billion in U.S. homes, according to the NPD group [2013]. Retrieved from http://www.prweb.com/releases/2013/3/prweb10542447.htm). A survey from 2013 found that Digital Natives switch between digital devices in nonworking hours every other minute (twenty-seven switches per hour), whereas Digital Immigrants switch seventeen times per hour (Moses, L. [March 31, 2013]. What does that second screen mean for viewers and advertisers? Retrieved from http://www.adweek.com/news/technology/what-does -second-screen-mean-viewers-and-advertisers-148240).

3. In 2010 U.S. youths ages eight to eighteen reported spending more than seven and a half hours a day in front of a screen watching TV, listening to music, surfing the Web, social networking, and playing videogames (Rideout, V. J., Foehr, U. G., and Roberts, D. F. [2010]. *Generation M2: Media in the lives of 8- to 18-year-olds.* Retrieved from http://kaiserfamilyfoundation .wordpress.com/uncategorized/report/generation-m2-media-in-the-lives-of

-8-to-18-year-olds). There was a significant jump from the group of eight-to-ten-year-olds, who spend an average of 7.51 hours in the cyber-world, to the older eleven-to-fourteen-year-olds, with an astonishing 11.53 hours, and fifteen-to-eighteen-year-olds, with a similar 11.23 hours. While TV watching still outstrips Internet use on average in adults (Pew Internet. [2012]. *Trend data [adults]*. Retrieved from http://pewinternet.org/ Trend-Data-%28Adults%29/Online-Activites-Total.aspx), in youths the rate of TV to Internet use for U.K. twelve-to-fifteen-year-olds in 2012 was evenly matched (17 hours on each activity per week) (Ofcom. [2012]. *Children and parents: Media use and attitudes report.* Retrieved from http://stakeholders. ofcom.org.uk/binaries/research/media-literacy/oct2012/main.pdf), whereas 2013 data for U.S. youths and adults shows that TV use is declining, with the biggest differences seen in twelve- to-twenty-four-year-olds, with their TV watching down three hours per week compared to 2011 data (Marketing Charts. [2013]. Are young people watching less TV? Retrieved from http:// www.marketingcharts.com/wp/television/are-young-people-watching-less -tv-24817). Furthermore, in 2012, for the first time in twenty years, the number of homes in the United States with TVs decreased (Stelter, B. [May 3, 2011]. Ownership of TV sets falls in U.S. *New York Times.* Retrieved from http://www .nytimes.com/2011/05/03/business/media/03television.html?_r=0&adxnnl=1 &ref=business&adxnnlx=1396530217-uFZGwm27zoGqpRHf4pOFog). Poverty is one reason given for this effect, in addition to the increasing number of youths who have been raised on laptops becoming young adults and starting their own households, for whom the computer offers all that a TV can, and more.

4. IDC. (2013). *Always connected: How smartphones and social keep us engaged.* Retrieved from https://fb-public.box.com/s/3iq5x6uwnqtq7ki4q8wk.
5. Rapoza, K. (February 18, 2013). One in five Americans work from home, numbers seen rising over 60 percent. *Forbes.* Retrieved from http://www .forbes.com/sites/kenrapoza/2013/02/18/one-in-five-americans-work-from -home-numbers-seen-rising-over-60.
6. Pew Internet, 2012.
7. Office for National Statistics. (2013). *Internet access—households and individuals, 2012 part 2.* Retrieved from http://www.ons.gov.uk/ons/dcp171778 _301822.pdf.
8. Entertainment Software Association, 2013.
9. Nielsen. (2011). *State of the media: The social media report.* Retrieved from http://cn.nielsen.com/documents/Nielsen-Social-Media-Report_FINAL _090911.pdf.
10. Bohannon, J. [June 6, 2013]. Online marriage is a happy marriage. Retrieved from http://www.smh.com.au/comment/online-marriage-is-a-happy-marriage -20130606-2ns0b.html.
11. Moss, S. (2010). *Natural childhood.* Retrieved from http://www.nationaltrust .org.uk/document-1355766991839.
12. Frost, J. L. (2010). *A history of children's play and play environments: Toward a contemporary child-saving movement.* New York: Routledge, p. 2.
13. Palmer, S. (2007). *Toxic childhood: How the modern world is damaging our children and what we can do about it.* London: Orion. The list is: 1. Climb a tree; 2. Roll down a really big hill; 3. Camp out in the wild; 4. Build a den; 5. Skim a stone; 6. Run around in the rain; 7. Fly a kite; 8. Catch a fish with a net;

9. Eat an apple straight from a tree; 10. Play conkers; 11. Throw some snow; 12. Hunt for treasure on the beach; 13. Make a mud pie; 14. Dam a stream; 15. Go sledging; 16. Bury someone in the sand; 17. Set up a snail race; 18. Balance on a fallen tree; 19. Swing on a rope swing; 20. Make a mud slide; 21. Eat blackberries growing in the wild; 22. Take a look inside a tree; 23. Visit an island; 24. Feel like you're flying in the wind; 25. Make a grass trumpet; 26. Hunt for fossils and bones; 27. Watch the sun wake up; 28. Climb a huge hill; 29. Get behind a waterfall; 30. Feed a bird from your hand; 31. Hunt for bugs; 32. Find some frogspawn; 33. Catch a butterfly in a net; 34. Track wild animals; 35. Discover what's in a pond; 36. Call an owl; 37. Check out the crazy creatures in a rock pool; 38. Bring up a butterfly; 39. Catch a crab; 40. Go on a nature walk at night; 41. Plant it, grow it, eat it; 42. Go wild swimming; 43. Go rafting; 44. Light a fire without matches; 45. Find your way with a map and compass; 46. Try bouldering; 47. Cook on a campfire; 48. Try abseiling; 49. Find a geocache; 50. Canoe down a river.

14. Moss, S. (2010). *Natural childhood*. Retrieved from http://www.nationaltrust .org.uk/document-1355766991839.
15. Moss, 2010, p. 6. Cited in Byron, T. (2008). Safer children in a digital world: the report of the Byron Review. Retrieved from http://media.education.gov .uk/assets/files/pdf/s/safer%20children%20in%20a%20digital%20world%20 the%202008%20byron%20review.pdf?

CHAPTER 3. A CONTROVERSIAL ISSUE

1. Byron, T. (2008). *Safer children in a digital world: The report of the Byron Review*. Retrieved from http://media.education.gov.uk/assets/files/pdf/s/safer%20 children%20in%20a%20digital%20world%20the%202008%20byron%20 review.pdf.
2. Howard-Jones, P. (2011). *The impact of digital technologies on human wellbeing: Evidence from the sciences of mind and brain*. Retrieved from http://www .nominettrust.org.uk/sites/default/files/NT%20SoA%20-%20The%20impact %20of%20digital%20technologies%20on%20human%20wellbeing.pdf, p. 5.
3. Rosen, L. D. (2012). *iDisorder: Understanding our obsession with technology and overcoming its hold on us*. New York: Macmillan.
4. Turkle, S. (2011). *Alone together: Why we expect more from technology and less from each other*. New York: Basic Books.
5. Batty, D. (February 24, 2012). Twitter co-founder says users shouldn't spend hours tweeting. Retrieved from http://www.theguardian.com/technology/ 2012/feb/23/twitter-cofounder-biz-stone-tweeting-unhealthy.
6. Schonfeld, E. (March 7, 2009). Eric Schmidt tells Charlie Rose Google is "unlikely" to buy Twitter and wants to turn phones into TVs. Retrieved from http://techcrunch.com/2009/03/07/eric-schmidt-tells-charlie-rose-google-is -unlikely-to-buy-twitter-and-wants-to-turn-phones-into-tvs.
7. Michael Rich, associate professor at Harvard Medical School, warns: "Their [Digital Natives'] brains are rewarded not for staying on task but for jumping to the next thing. The worry is we're raising a generation of kids in front of screens whose brains are going to be wired differently" (Richtel, M. [November 21, 2010]. Growing up digital, wired for distraction. Retrieved from

http://www.nytimes.com/2010/11/21/technology/21brain.html?pagewanted
=all).

Jordan Grafman, chief of cognitive science at the National Institute of Neurological Disorders and Stroke, says: "In general, technology can be good (for children's cognitive development) if it is used judiciously. But if it is used in a non-judicious fashion, it will shape the brain in what I think will actually be a negative way . . . a lot of what is appealing about all these types of instant communications is that they are fast. Fast is not equated with deliberation. So I think they can produce a tendency toward shallow thinking. It's not going to turn off the brain to thinking deeply and thoughtfully about things, but it is going to make that a little bit more difficult to do" (Whitman, A. and Goldberg, J. [2008]. *Brain development in a hyper-tech world*. Retrieved from http://www.dana.org/media/detail.aspx?id=13126).

The American Academy of Pediatrics has suggested that two hours or more per day of computer use increases the probability of emotional, social, and attention problems, a view borne out in findings reported recently by Angie Page and colleagues at Bristol University, who concluded that children's screen viewing is related to psychological difficulties irrespective of physical activity. Participants were 1,013 children with an average age of almost eleven years, who self-reported average daily television hours and computer use on a questionnaire. Page found that greater television and computer use were related to higher psychological difficulty scores. Children who spent more than two hours per day watching television or using a computer—which would appear to be the majority of U.K. and U.S. children—were at increased risk of high levels of psychological difficulties, and this risk increased if the children also failed to meet physical activity guidelines (Page, A. S., Cooper, A. R., Griew, P., and Jago, R. [2010]. Children's screen viewing is related to psychological difficulties irrespective of physical activity. *Pediatrics* 126, no. 5, e1011–e1017. doi:10:1542/peds.2010-1154).

Michael Friedlander, head of neuroscience at Baylor College of Medicine, has said: "If a child is doing homework while on the computer engaged in chat rooms, or listening to iTunes and so forth, I do think there is a risk that there will never be enough depth and time spent on any one component to go as deep or as far as you might have. You might satisfactorily get all these things done, but the quality of the work or of the communication may not reach the level that it could have had it been given one's full attention. There's a risk of being a mile wide and an inch deep" (Whitman and Goldberg, 2008: see above).

8. Bavelier, D., Green, C. S., Han, D. H., Renshaw, P. F., Merzenich, M. M., and Gentile, D. A. (2011). Brains on videogames. *Nature Reviews Neuroscience* 12, no. 12, 763–768. doi:10:1038/nrn3135, p. 766.

9. Pearson UK. (2012). New "Enjoy Reading" campaign and support materials launched to help parents and teachers switch children on to reading for life. Retrieved from http://uk.pearson.com/home/news/2012/october/new-_enjoy-reading-campaign-and-support-materials-launched-to-he.html.

10. Purcell, K., Rainie, L., Heaps, A., Buchanan, J., Friedrich, L., Jacklin, A., . . . and Zickuhr, K. (2012). *How teens do research in the digital world*. Retrieved from http://www.pewinternet.org/Reports/2012/Student-Research.aspx, p. 2.

11. Those signing this statement were a diverse bunch, from household names like best-selling children's author Philip Pullman, to influential psycholo-

gist Oliver James, as well as the founder of Kids' Company, the charity for the homeless young, Camilla Batmanghelidjh. The diversity of sectors represented certainly revealed the sweep of issues involved—after all, lifestyle is hardly a single activity or issue that is the monopoly of any one narrow field of expertise (Erosion of childhood: Letter with full list of signatories. [September 23, 2011]. Retrieved from http://www.telegraph.co.uk/education/educationnews/8784996/Erosion-of-childhood-letter-with-full-list-of-signatories.html).

12. Anderson, J. Q., and Rainie, L. (2012). *Millennials will benefit and suffer due to their hyperconnected lives*. Retrieved from http://www.elon.edu/docs/e-web/predictions/expertsurveys/2012survey/PIP_Future_of_Internet_2012_Gen_Always_ON.pdf.

13. Vinter, P. (September 1, 2012). Zadie Smith pays tribute to computer software that blocks Internet sites allowing her to write new book without distractions. Retrieved from http://www.dailymail.co.uk/news/article-2196718/Zadie-Smith-pays-tribute-software-BLOCKS-internet-sites-allowing-write-new-book-distractions.htm.

14. World Economic Forum. (2013). *Global risks report 2013* (8th ed.). Retrieved from http://reports.weforum.org/global-risks-2013, pp. 23–24. The report states: "The Internet remains an uncharted, fast-evolving territory. Current generations are able to communicate and share information instantaneously and at a scale larger than ever before. Social media increasingly allows information to spread around the world at breakneck speed. While the benefits of this are obvious and well documented, our hyperconnected world could also enable the rapid viral spread of information that is either intentionally or unintentionally misleading or provocative, with serious consequences . . . It is just as conceivable that the offending content's original author might not even be aware of its misuse or misrepresentation by others on the Internet, or that it was triggered by an error in translation from one language to another. We can think of such a scenario as an example of a digital wildfire." Such an example occurred in 2012, when someone impersonating a Russian parliamentarian tweeted that Syrian president Bashar al-Assad had been killed or injured. Crude oil prices rose in response before the tweet was revealed to be a hoax (Howell, L. [January 8, 2013]. Only you can prevent digital wildfires. Retrieved from http://www.nytimes.com/2013/01/09/opinion/only-you-can-prevent-digital-wildfires.html?_r=0).

15. Greenfield, S. (February 12, 2009). Children: Social networking sites. U.K. Parliament, House of Lords. Retrieved from http://www.publications.parliament.uk/pa/ld200809/ldhansrd/text/90212-0010.htm.

16. Ivo Quaritiroli in *The Digitally Divided Self* (http://www.amazon.com/The-Digitally-Divided-Self-Relinquishing/dp/8897233007) claims that "statements such as 'it is not scientific' or 'we don't have enough data' are typical defenses that technologically orientated people use to counteract criticism or expressions of concern" (Chapter 1, section "Technology can't be challenged").

17. A paradigm is, in Kuhn's own words, "what members of a scientific community, and they alone, share." According to Kuhn, a paradigm is more than just a single simple theory but the entire worldview within which it exists. Needless to say, such a view may encompass uncomfortable anomalies, facts, and findings that just don't fit, but which are for a while brushed aside

because of the intellectual discomfort they bring and also because of the explanatory void that might consequently yawn open. But as such anomalies, inevitably those from experimental data, start to accumulate, so some scientists may begin to doubt the whole perspective, not least because they have a more attractive new alternative that can encompass and account for all the erstwhile uncomfortable findings. A "crisis" ensues in the respective disciplines, so that eventually, as in France in 1789 and in Russia a little over a century later, a revolution takes place, a struggle between the old order and a new. Comparing such sweeping ideological struggles with academic wrangling might seem far-fetched, but it actually isn't that way off the mark. Bear in mind that what Kuhn was describing were completely different ways of seeing things, so radical that they would influence the way scientists, and therefore eventually everyone, saw the world for generations to come (Kuhn, T. S. [1977]. *The essential tension: Selected studies in scientific tradition and change*. Chicago: University of Chicago Press, p. 294).

18. Beattie-Moss, M. (February 4, 2008). Gut instincts: A profile of Nobel laureate Barry Marshall. Retrieved from http://news.psu.edu/story/140921/2008/02/04/research/gut-instincts-profile-nobel-laureate-barry-marshall.

19. The difficulty with the attitude that we cannot even talk about the prospects and implications for humanity of cyberculture until there is conclusive "scientific evidence" that it is either "good" or "bad" is well articulated by Dr. Aric Sigman, of the Royal Society of Medicine: "It strikes me as a terrible shame that our society requires photos of brains shrinking in order to take seriously the common-sense assumption that long hours in front of screens is not good for our children's health" (Harris, S. [July 18, 2011]. Too much Internet use "can damage teenagers' brains." Retrieved from http://www.dailymail.co.uk/sciencetech/article-2015196/Too-internet-use-damage-teenagers-brains.html).

20. Statistical analysis is conducted on research findings to determine whether the results of the study are likely to apply to the whole population in which the researchers are interested, beyond just the sample obtained for the study. When results of a study are statistically significant, it means that the findings, often in the form of a relationship between variables or a difference between groups of participants, are not likely to be due to chance. The conclusions drawn from statistical methods are sensitive to the particulars of the study design, including the selection of variables and the size of the sample examined. For example, a large sample yields high statistical power, which means that relatively small differences may be detected as statistically significant. Researchers must use their understanding of statistics and the subject matter to determine which of these findings are important as opposed to spurious. There is no magic rule regarding what size sample or number of participants is "large enough," and this choice in experimental design is somewhat arbitrary. Statistics do not provide an answer and researchers must make a choice based on their understanding of the variables of interest and the effect sizes they might anticipate. Additionally, statistical analysis cannot account for poor study design, such as how the participants were recruited or how the data collection process occurred. This means that if aspects of the study design were biased, this will increase the likelihood of finding a significant result. Moreover, researchers themselves can manipulate statistical analysis and the subsequent interpretation of results, as publi-

cation in a journal can often depend on finding a statistically significant result. Where appropriate, *Mind Change* will comment on significant study findings that may be biased in some way. However, it is beyond the scope of this book to go into this in too extensive detail.

CHAPTER 4. A MULTIFACETED PHENOMENON

1. Baede, A. P. M. (n.d.). Working Group I: The scientific basis. Intergovernmental Panel on Climate Change. Retrieved from http://www.ipcc.ch/ipccreports/tar/wg1/518.htm.
2. Witness the popularity of sites like Klout, which gives you a score for your importance in the cyberworld. Interestingly enough, "coolness" is now democratized: wealth, gender, and age are no longer relevant, but then neither is anything special that you may have done. So the interesting and unprecedented feature of being cool and famous on social networking sites is that such content need have nothing to do with your particular prowess in any area, and indeed nothing to do with the "real" you at all. It is important to bear in mind that the interaction between the brain and the environment is a two-way dialogue: just as vital to how we view and use the latest technology is the impact that an environment dominated by compulsive engagement with social networking sites will have on shaping our relationships and our view of our own identity.
3. Lenhart, A., Madden, M., Smith, A., Purcell, K., Zickuhr, K., and Rainie, L. (2011). *Teens, kindness and cruelty on social network sites.* Retrieved from http://pewinternet.org/Reports/2011/Teens-and-social-media.aspx, p. 28.
4. Konrath, S. H., O'Brien, E. H., and Hsing, C. (2011). Changes in dispositional empathy in American college students over time: A meta-analysis. *Personality and Social Psychology Review* 15, no. 2, 180–198. doi:10:1177/1088868310377395.
5. PR Newswire. (2013). Facebook reports first quarter 2013 result. Retrieved from http://www.prnewswire.com/news-releases/205652631.html.
6. Internet World Stats (2012). Facebook users in the world: Facebook usage and Facebook growth statistics by world geographic regions. Retrieved from http://www.internetworldstats.com/facebook.htm.
7. Twitter. (December 18, 2012). There are now more than 200M monthly active @twitter users. You are the pulse of the planet. We're grateful for your ongoing support! [Twitter post]. Retrieved from https://twitter.com/twitter/status/281051652235087872.
8. Ofcom. (2013). *Adults' media use and attitudes report.* Retrieved from http://stakeholders.ofcom.org.uk/binaries/research/media-literacy/adult-media-lit-13/2013_Adult_ML_Tracker.pdf.
9. Madden, M., Lenhart, A., Duggan, M., Cortesi, S., and Gasser, U. (2013). *Teens and technology 2013.* Retrieved from http://pewinternet.org/~/media//Files/Reports/2013/PIP_TeensandTechnology2013.pdf.
10. Arbitron Inc. and Edison Research. (2013). *The infinite dial 2013: Navigating digital platforms.* Retrieved from http://www.edisonresearch.com/wp-content/uploads/2013/04/Edison_Research_Arbitron_Infinite_Dial_2013.pdf.
11. Smith, C. (2013). By the numbers: 32 amazing Facebook stats [blog post, updated June 2013]. Retrieved from http://expandedramblings.com/index.php/by-the-numbers-17-amazing-facebook-stats.

12. Arbitron Inc. and Edison Research, 2013.
13. Hampton, K. N., Goulet, L. S., Rainie, L., and Purcell, K. (2011). *Social networking sites and our lives*. Retrieved from http://pewinternet.org/Reports/2011/Technology-and-social-networks.aspx.
14. Hampton et al., 2011.
15. McAfee. (2010). *The secret online lives of teens*. Retrieved from http://us.mcafee.com/en-us/local/docs/lives_of_teens.pdf.
16. Government Office for Science, London. (2013). *Foresight future identities: Final project report*. Retrieved from http://www.bis.gov.uk/foresight/our-work/policy-futures/identity.
17. Gentile, D. A., and Anderson, C. A. (2003). Violent videogames: The newest media violence hazard. In D. A. Gentile (Ed.), *Media violence and children: A complete guide for parents and professionals* (Vol. 22). Retrieved from www.psychology.iastate.edu/faculty/caa/abstracts/2000-2004/03GA.pdf.
18. By 2005, a national study commissioned by U.K. Games Research of individuals between six and sixty-five years of age showed a clear age factor slanted toward youth: more than 80 percent of those under the age of twenty-four were playing videogames (Pratchett, R. [2005]. *Gamers in the UK: Digital play, digital lifestyles*. Retrieved from http://crystaltips.typepad.com/wonderland/files/bbc_uk_games_research_2005.pdf). In 2008, 97 percent of American teens were playing videogames (Lenhart, A., Jones, S., and Macgill, A. R. [2008] *Adults and videogames*. Retrieved from http://www.pewinternet.org/Reports/2008/Adults-and-Video-Games/1-Data-Memo.aspx), while within a few years (2011) in Australia the number was similar, 94 percent (Digital Australia. [2011]. *Key findings*. Retrieved from http://www.igea.net/wp-content/uploads/2011/10/DA12KeyFindings.pdf). Although these statistics come from different countries, the English-speaking developed world cultures are surely similar enough for a comparable trend and trajectory to be discerned.
19. Homer, B. D., Hayward, E. O., Frye, J., and Plass, J. L. (2012). Gender and player characteristics in videogame play of preadolescents. *Computers in Human Behavior* 28, no. 5, 1782–1789. doi:10:1016/j.chb.2012:04.018.
20. Rideout, V. J., Foehr, U. G., and Roberts, D. F. (2010). *Generation M2: Media in the lives of 8- to 18-year-olds*. Retrieved from http://kaiserfamilyfoundation.wordpress.com/uncategorized/report/generation-m2-media-in-the-lives-of-8-to-18-year-olds.
21. Cummings, H. M., and Vandewater, E. A. (2007). Relation of adolescent videogame play to time spent in other activities. *Archives of Pediatrics & Adolescent Medicine* 161, no. 7, 684. doi:10:1001/archpedi.161:7.684.
22. Cooper, R. (February 3, 2012). Gamer lies dead in Internet café for 9 hours before anyone notices. Retrieved from http://www.dailymail.co.uk/news/article-2096128/Gamer-lies-dead-Taiwan-internet-cafe-9-HOURS-notices.html.
23. Diablo 3 death: Teen dies after playing game for 40 hours straight. (July 18, 2012). Retrieved from http://www.huffingtonpost.com/2012/07/18/diablo-3-death-chuang-taiwan-_n_1683036.html.
24. Tran, M. (March 6, 2010). Girl starved to death while parents raised virtual child in online game. Retrieved from http://www.theguardian.com/world/2010/mar/05/korean-girl-starved-online-game.
25. Carter, H. (November 19, 2010). Man jailed for murder of girlfriend's toddler. Retrieved from http://www.theguardian.com/uk/2010/nov/18/man-jailed-murder-girlfriends-toddler.

26. Videogame fanatic hunts down and stabs rival player who killed character online. (May 27, 2010). Retrieved from http://www.telegraph.co.uk/news/worldnews/europe/france/7771505/Video-game-fanatic-hunts-down-and-stabs-rival-player-who-killed-character-online.html.

27. Anderson, C. A., Shibuya, A., Ihori, N., Swing, E. L., Bushman, B. J., Sakamoto, A., . . . and Saleem, M. (2010). Violent videogame effects on aggression, empathy, and prosocial behavior in Eastern and Western countries: A meta-analytic review. *Psychological Bulletin* 136, no. 2, 151. doi:10:1037/a0018251.

28. Kühn, S., Romanowski, A., Schilling, C., Lorenz, R., Mörsen, C., Seiferth, N., . . . and Gallinat, J. (2011). The neural basis of gaming. *Translational Psychiatry* 1, no. 11, e53. doi:10:1038/tp.2011:53.

29. Sullivan, D. (February 11, 2013). Google still world's most popular search engine by far, but share of unique searchers dips slightly. Retrieved from http://searchengineland.com/google-worlds-most-popular-search-engine-148089.

30. Mangen, A., Walgermo, B. R., and Brønnick, K. (2013). Reading linear texts on paper versus computer screen: Effects on reading comprehension. *International Journal of Educational Research* 58, 61–68. doi:10:1016/j.ijer.2012:12.002.

CHAPTER 5. HOW THE BRAIN WORKS

1. The brainstem is the extension of the spinal cord that forms the inner core of the brain, around which other structures are elaborated. This is functionally the most basic part of the brain, shared even with reptiles. It mediates respiration, sleep-wake cycles, and arousal. Of many possible reviews, see: Siegel, J. (2004). Brain mechanisms that control sleep and waking. *Naturwissenschaften* 91, no. 8, 355–65; Jones, B. E. (2003). Arousal systems. *Frontiers in Bioscience* 8, 438–451.

2. The cerebellum: nicknamed the "autopilot" of the brain and mediating fine-tuned sensorimotor coordination. For a recent review, see Reeber, S. L., Otis, T. S., and Sillitoe, R. V. (2013). New roles for the cerebellum in health and disease. *Frontiers in Systems Neuroscience* 7, 83.

3. The cortex: unlike the brainstem and the cerebellum, this is a newer, indeed the newest, brain region in terms of evolution. It is organized in repeating modular circuits like a cookie cutter. Some areas are related to a single sense, while others serve more "cognitive" functions like learning and memory and are referred to by the umbrella term "association cortex." See Shipp, S. (2007). Structure and function of the cerebral cortex. *Current Biology* 17, 443–449.

4. This blip is more precisely an "action potential": there is a sharp change in the potential difference (voltage) across the cell membrane caused by positively charged sodium ions rushing into the cell, making it depolarized, a situation that then triggers the efflux of positively charged potassium ions, once again making the potential difference more negative. For more detailed descriptions, see Purves, D., Augustine, G. J., Fitzpatrick, D., Hall, W. C., La-Mantia, A. S., and White, L. E. (Eds.) (2012). *Neuroscience* (5th ed.). Sunderland, MA: Sinauer.

5. The "nerve terminal" is the end of the axon, the long process emanating from the cell body along which the action potential is propagated at several

hundred miles per hour. Once the "blip" invades the terminal, the change in voltage triggers the emptying of small packets (vesicles) containing neurotransmitter into the synaptic cleft.

6. Purves et al. (2012).

7. For example, it could be the case that the input from one neuron "A" caused a small depolarization, but not large enough to bring the voltage of the cell to the threshold for being able to generate a full-blown action potential. Now imagine that, during this time period while the voltage was raised, another input "B" arrived that also on its own would normally have caused only a subthreshold depolarization: because A + B could summate to threshold within this time window, an action potential could now occur that would not have been possible if the two inputs had not arrived relatively close together.

8. "Modulation": a term used when a neurotransmitter or other bioactive compound has no effect on its own but enhances or diminishes the action of another signaling molecule.

9. The most familiar and easiest way of thinking about brain organization is as a hierarchy, similar to a chain of command with the boss at the top of a pyramid-like structure. Indeed this concept fitted well with scientific findings in the 1960s when two physiologists, David Hubel and Torsten Weisel, made a breakthrough, Nobel Prize–winning discovery. Hubel and Weisel were working on the visual system and monitoring the activity of single brain cells in the different brain regions which processed inputs from the retina, and then further on into the depths of the brain. Their remarkable finding was that, as they went deeper into the brain, further away from the initial processing of the retina, the cells seemed to become fussier in terms of what turned them on, literally. Initially, the sight of any old blob would excite a neuron, but further up the chain of command it might have to be a line, and then only a line in a certain orientation, and then a line in a certain orientation but only moving in a specific direction (Hubel, D. H., and Weisel, T. N. [1962]. Receptive fields, binocular interaction and functional architecture in the cat's visual cortex. *Journal of Physiology* 160, no. 1, 106–154. Retrieved from http://www.ncbi.nlm.nih.gov/pmc/articles/PMC1359523/pdf/jphysiol01247-0121.pdf). It was certainly an amazing discovery that a single brain cell could have such an individual signature, but it led to some strange extrapolations. You can see how Hubel and Weisel's discovery easily led to the notion that the further up the hierarchy of the brain you went, the fussier the cell would become, eventually responding only to very sophisticated images, such as a face or even a specific face. The terminology of the time liked to refer to a hypothetical "grandmother cell" which, as its name suggested, would only respond to the sight of your grandmother as the ultimate stage in the organization. Although, much more recently, Christof Koch and his team of researchers at Caltech recorded cells in the brains of conscious neurosurgical patients specifically responding, for example, to pictures of Halle Berry (Quiroga, R. Q., Reddy, L., Kreiman, G., Koch, C., and Fried, I. [2005]. Invariant visual representation by single neurons in the human brain. *Nature* 435, no. 7045, 1102–1107. doi:10:1038/nature03687), the idea that a single "Berry cell" or a grandmother cell could effectively be "the boss" has been largely discredited, if only by simple logic. If you never had a grandmother, a cell would be wasted; or if you did have a grandmother but

your grandmother cell died, as many neurons do daily, then you'd never recognize your grandmother ever again! Just as a brain region can't be an independent "center," it is even less likely that a single brain cell can be a final destination—and it certainly can't be the ultimate boss. What would "the boss" do subsequently? After all, there would be no one further to instruct.

10. Kolb, B. (2009). Brain and behavioral plasticity in the developing brain: Neuroscience and public policy. *Paediatrics & Child Health* 14, no. 10, 651–652. Retrieved from http://www.ncbi.nlm.nih.gov/pmc/articles/PMC2807801.

CHAPTER 6. HOW THE BRAIN CHANGES

1. Maguire, E. A., Gadian, D. G., Johnsrude, I. S., Good, C. D., Ashburner, J., Frackowiak, R. S., and Frith, C. D. (2000). Navigation-related structural change in the hippocampi of taxi drivers. *Proceedings of the National Academy of Sciences* 97, no. 8, 4398–4403. doi:10:1073/pnas.070039597.

2. The octopus, who featured in classic memory experiments in the 1960s and, more recently, received much attention when one of their number, "Paul," showed apparently prescient powers in being able to predict the outcomes of various matches in the 2011 World Cup. See also Young, J. Z. (1983). The distributed tactile memory system of Octopus. *Proceedings of the Royal Society of London. Series B, Biological Sciences,* 135–176.

3. Abrams, T. W., and Kandel, E. R. (1988). Is contiguity detection in classical conditioning a system or a cellular property? Learning in *Aplysia* suggests a possible molecular site. *Trends in Neurosciences* 11, no. 4, 128–135. doi:10:1016/0166-2236(88)90137-3.

4. Doidge, N. (2007). *The brain that changes itself: Stories of personal triumph from the frontiers of brain science.* New York: Penguin, p. 315.

5. Rosenzweig, M. R. (1996). Aspects of the search for neural mechanisms of memory. *Annual Review of Psychology* 47, no. 1, 1–32. doi:10:1146/annurev.psych.47:1.1. Rosenzweig, M. R., Modification of Brain Circuits through Experience, in Neural Plasticity and Memory: From Genes to Brain Imaging, F. Bermúdez-Rattoni, (Ed.), 2007, CRC Press: Boca Raton, Florida, USA.

6. White House, Office of the Press Secretary. (April 17, 1997). Remarks from the Conference. Retrieved from http://clinton4.nara.gov/WH/New/ECDC/Remarks.html.

7. Bavelier, D., and Neville, H. J. (2002). Cross-modal plasticity: Where and how? *Nature Reviews Neuroscience* 3, no. 6, 443–452. doi:10:1038/nrn848.

8. Derbyshire, D. (March 6, 2011). The boy whose damaged brain "rewired" itself. Retrieved from http://www.telegraph.co.uk/news/uknews/1325183/The-boy-whose-damaged-brain-rewired-itself.html.

9. Lewis, T. L., and Maurer, D. (2005). Multiple sensitive periods in human visual development: Evidence from visually deprived children. *Developmental Psychobiology* 46, no. 3, 163–183. doi:10:1002/dev.20055.

10. Neville, H. J., and Lawson, D. (1987). Attention to central and peripheral visual space in a movement detection task: An event-related potential and behavioral study. II. Congenitally deaf adults. *Brain Research* 405, no. 2, 268–283. doi:10:1016/0006-8993(87)90296-4.

11. Kleim, J. A. (2011). Neural plasticity and neurorehabilitation: Teaching the new brain old tricks. *Journal of Communication Disorders* 44, no. 5, 521–528. doi:10:1016/j.jcomdis.2011:04.006.

12. Schlaug, G., Marchina, S., and Norton, A. (2009). Evidence for plasticity in white-matter tracts of patients with chronic Broca's aphasia undergoing intense intonation-based speech therapy. *Annals of the New York Academy of Sciences* 1169, no. 1, 385–394. doi:10:1111/j.1749-6632:2009:04587.x.

13. Nudo, R. J. (2011). Neural bases of recovery after brain injury. *Journal of Communication Disorders* 44, no. 5, 515–520. doi:10:1016/j.jcomdis.2011:04.004.

14. Where did such a bizarre idea come from? One suggestion is that the great psychologist William James was working with an accelerated program of learning for a child prodigy in the 1890s, and generalized from this exceptional case that most people only realized a fraction of their true potential. Maybe so, but it's not because 90 percent of our brains are not working. This strangely precise figure has been attributed to the American writer Lowell Thomas who, in 1936, tried to summarize James's work. Perhaps he based the estimate on the percentage of brain functions that could be mapped at the time in terms of brain location. While Thomas may not have been party to our current knowledge of the brain, 90/10 is a ratio that coincidentally still features. For example, the key nerve cells of the brain, neurons, are outnumbered ten to one by glial cells (the name is from the Greek for "glue"), which take care of the basic cerebral housekeeping and ensure a healthy and nurturing brain environment. Moreover, at any one time, only about 10 percent of neurons are spontaneously active. However, this isn't to say that the remainder are dead or inactive, any more than a soccer player standing alert but briefly stationary on the field would be regarded as not participating in the game.

15. Jenkins, W. M., Merzenich, M. M., Ochs, M. T., Allard, T., and Guic-Robles, E. (1990). Functional reorganization of primary somatosensory cortex in adult owl monkeys after behaviorally controlled tactile stimulation. *Journal of Neurophysiology* 63, no. 1, 82–104. Retrieved from http://jn.physiology.org/content/63/1/82.full.pdf+html.

16. Elbert, T., Pantev, C., Wienbruch, C., Rockstroh, B., and Taub, E. (1995). Increased cortical representation of the fingers of the left hand in string players. *Science* 270, no. 5234, 305–307. doi:10:1126/science.270:5234:305.

17. Gaser, C., and Schlaug, G. (2003). Brain structures differ between musicians and non-musicians. *Journal of Neuroscience* 23, no. 27, 9240–9245. Retrieved from http://www.jneurosci.org/content/23/27/9240.full.pdf+html.

18. Aydin, K., Ucar, A., Oguz, K. K., Okur, O. O., Agayev, A., Unal, Z., . . . and Ozturk, C. (2007). Increased gray matter density in the parietal cortex of mathematicians: A voxel-based morphometry study. *American Journal of Neuroradiology* 28, no. 10, 1859–1864. doi:10:3174/ajnr.A0696.

19. Park, I. S., Lee, K. J., Han, J. W., Lee, N. J., Lee, W. T., and Park, K. A. (2009). Experience-dependent plasticity of cerebellar vermis in basketball players. *The Cerebellum* 8, no. 3, 334–339. doi:10:1007/s12311-009-0100-1.

20. Jäncke, L., Koeneke, S., Hoppe, A., Rominger, C., and Hänggi, J. (2009). The architecture of the golfer's brain. *PLOS ONE* 4, no. 3, e4785. doi:10:1371/journal.pone.0004785.

21. Draganski, B., Gaser, C., Busch, V., Schuierer, G., Bogdahn, U., and May, A. (2004). Neuroplasticity: Changes in gray matter induced by training. *Nature*

427, no. 6972, 311–312. doi:10:1038/427311a. Driemeyer, J., Boyke, J., Gaser, C., Büchel, C., and May, A. (2008). Changes in gray matter induced by learning: Revisited. *PLOS ONE* 3, no. 7, e2669. doi:10:1371/journal.pone.0002669.

22. Boyke, J., Driemeyer, J., Gaser, C., Büchel, C., and May, A. (2008). Training-induced brain structure changes in the elderly. *Journal of Neuroscience* 28, no. 28, 7031–7035. doi:10:1523/JNEUROSCI.0742-08:2008.

23. Engvig, A., Fjell, A. M., Westlye, L. T., Moberget, T., Sundseth, Ø., Larsen, V. A., and Walhovd, K. B. (2010). Effects of memory training on cortical thickness in the elderly. *Neuroimage* 52, no. 4, 1667–1676. doi:10:1016/j.neuroimage.2010:05.041.

24. Draganski, B., Gaser, C., Kempermann, G., Kuhn, H. G., Winkler, J., Büchel, C., and May, A. (2006). Temporal and spatial dynamics of brain structure changes during extensive learning. *Journal of Neuroscience* 26, no. 23, 6314–6317. doi:10:1523/JNEUROSCI.4628-05:2006.

25. May, A. (2011). Experience-dependent structural plasticity in the adult human brain. *Trends in Cognitive Sciences* 15, no. 10, 475–482. doi:10:1016/j.tics.2011:08.002, p. 4.

26. Mechelli, A., Crinion, J. T., Noppeney, U., O'Doherty, J., Ashburner, J., Frackowiak, R. S., and Price, C. J. (2004). Neurolinguistics: Structural plasticity in the bilingual brain. *Nature* 431, no. 7010, 757. doi:10:1038/431757a. Stein, M., Federspiel, A., Koenig, T., Wirth, M., Strik, W., Wiest, R., . . . and Dierks, T. (2012). Structural plasticity in the language system related to increased second language proficiency. *Cortex* 48, no. 4, 458–465. doi:10:1016/j.cortex.2010:10.007.

27. Begley, S. (2008). *The plastic mind*. London: Constable & Robinson.

28. Pickren, W., and Rutherford, A. (2010). *A history of modern psychology in context*. Hoboken, NJ: Wiley.

29. Diamond, M. C., Krech, D., and Rosenzweig, M. R. (1964). The effects of an enriched environment on the histology of the rat cerebral cortex. *Journal of Comparative Neurology* 123, no. 1, 111–119. doi:10:1002/cne.901230110.

30. Van Dellen, A., Blakemore, C., Deacon, R., York, D., and Hannan, A. J. (2000). Delaying the onset of Huntington's in mice. *Nature* 404, no. 6779, 721–722. doi:10:1038/35008142.

31. Amaral, O. B., Vargas, R. S., Hansel, G., Izquierdo, I., and Souza, D. O. (2008). Duration of environmental enrichment influences the magnitude and persistence of its behavioral effects on mice. *Physiology & Behavior* 93, no. 1, 388–394. doi:10:1016/j.physbeh.2007:09.009.

32. Johansson, B. B. (1996). Functional outcome in rats transferred to an enriched environment 15 days after focal brain ischemia. *Stroke* 27, no. 2, 324–326. doi:10:1161/01.STR.27:2.324.

33. Young, D., Lawlor, P. A., Leone, P., Dragunow, M., and During, M. J. (1999). Environmental enrichment inhibits spontaneous apoptosis, prevents seizures and is neuroprotective. *Nature Medicine* 5, no. 4, 448–453. doi:10:1038/7449.

34. Mohammed, A. H., Zhu, S. W., Darmopil, S., Hjerling-Leffler, J., Ernfors, P., Winblad, B., . . . and Bogdanovic, N. (2002). Environmental enrichment and the brain. *Progress in Brain Research* no. 138, 109–133. doi:10:1016/S0079-6123(02)38074-9, p. 127.

35. Hebb, D. O. (1949). *The organization of behavior: A neuropsychological theory*. New York: Wiley.

36. But this simple chain of electrical and chemical events will not explain how synapses can become "stronger" (more efficient and effective) the more they are used: something additional must be happening to cause such plasticity. Bliss and Lomo's great discovery was to show that some of the target molecules (receptors) on the receiving cell can be quite fussy about the conditions in which they will work well, and this fussiness can be turned to advantage and form the basis for the adaptability of brain cells. For the fussy type of receptor a simple handshake is not enough, even when interlocking with a neurotransmitter X; it is just not sufficient to actually cause a change in voltage in the cell. Or, to use another analogy, the boat may be in dock but no car is yet available. Something else must happen next; there must be a further change while neurotransmitter X is already present. The handshake will be effective not just because two hands interlock, but because one of them now squeezes the other. Accordingly, if a second neurotransmitter, Y, now arrives on the scene and also docks into the cell, the contingency of X and Y will at last fulfill the demands of the fussy receptor (a car will appear). An electrical signal will now be generated, but with longer-term consequences. When the fussy receptor starts to work, it will trigger the opening of little channels in the cell so that calcium can flood in. In turn, the calcium will release a chemical that returns back across the synapse to the original incoming cell and makes it release yet still more neurotransmitter than usual. Meanwhile, within the target cell, a cascade of events is set in train that will make the cell more sensitive in terms of how effectively it will respond to the standard amount of incoming input. The same signal will have a much more powerful effect. The synapse now works more powerfully, but things don't just stop there. The calcium that has entered the cell during this process (long-term potentiation) has still longer-term actions: more specialized chemicals are produced inside the cell that stabilize the synapse still further by acting like sticky badges (cell adhesion molecules). Meanwhile, different proteins kick in to enhance the appearance of yet more neuronal contacts. All this has happened because of the initial requirement of the fussy receptor where two things had to happen within a certain time frame, and for a sustained period of time, in order for calcium to infiltrate the neuron. In this way, the more a behavior is repeated or rehearsed, such as a repetitious response to a certain experience, the greater the effect and the stronger the respective synapses will become over time, and hence that experience will literally leave its mark on the brain.

37. Scarmeas, N. and Stern, Y. (2003). Cognitive reserve and lifestyle. *Journal of Clinical and Experimental Neuropsychology* 25, no. 5, 625–633. doi:10:1076/jcen.25:5.625:14576.

38. Frasca, D., Tomaszczyk, J., McFadyen, B. J., and Green, R. E. (2013). Traumatic brain injury and post-acute decline: What role does environmental enrichment play? A scoping review. *Frontiers in Human Neuroscience* no. 7, 31. doi:10:3389/fnhum.2013:00031.

39. Scarmeas and Stern, 2003.

40. Frasca et al., 2013.

41. Winocur, G., and Moscovitch, M. (1990). A comparison of cognitive function in community-dwelling and institutionalized old people of normal intelligence. *Canadian Journal of Psychology/Revue Canadienne de Psychologie* 44, no. 4, 435–444. doi:10:1037/h0084270.

42. Olson, A. K., Eadie, B. D., Ernst, C., and Christie, B. R. (2006). Environmental enrichment and voluntary exercise massively increase neurogenesis in the adult hippocampus via dissociable pathways. *Hippocampus* 16, no. 3, 250–260. doi:10:1002/hipo.20157.

43. Nottebohm, F. (2002). Neuronal replacement in adult brain. *Brain Research Bulletin* 57, no. 6, 737–749. doi:10:1016/S0361-9230(02)00750-5.

44. Nyberg, L., Lövdén, M., Riklund, K., Lindenberger, U., and Bäckman, L. (2012). Memory aging and brain maintenance. *Trends in Cognitive Sciences* 16, no. 5, 292–305. doi:10:1016/j.tics.2012:04.005.

45. Mu, Y. and Gage, F. H. (2011). Adult hippocampal neurogenesis and its role in Alzheimer's disease. *Molecular Neurodegeneration* 6, no. 1, 85. doi:10:1186/1750-1326-6-85.

46. Pascual-Leone, A., Nguyet, D., Cohen, L. G., Brasil-Neto, J. P., Cammarota, A., and Hallett, M. (1995). Modulation of muscle responses evoked by transcranial magnetic stimulation during the acquisition of new fine motor skills. *Journal of Neurophysiology* 74, no. 3, 1037–1045. Retrieved from http://psycnet.apa.org/psycinfo/1996-25629-001.

47. Van Praag, H., Kempermann, G., and Gage, F. H. (1999). Running increases cell proliferation and neurogenesis in the adult mouse dentate gyrus. *Nature Neuroscience* 2, no. 3, 266–270. doi:10:1038/6368.

48. Sauro, M. D., and Greenberg, R. P. (2005). Endogenous opiates and the placebo effect: A meta-analytic review. *Journal of Psychosomatic Research* 58, 115–20.

49. Tanti, A. and Belzung, C. (2013). Hippocampal neurogenesis: A biomarker for depression or antidepressant effects? Methodological considerations and perspectives for future research. *Cell and Tissue Research,* 1–17. doi:10:1007/s00441-013-1612-z.

50. Begley, 2008.

CHAPTER 7. HOW THE BRAIN BECOMES A MIND

1. Penfield, W., and Boldrey, E. (1937). Somatic motor and sensory representation in the cerebral cortex of man as studied by electrical stimulation. *Brain: A Journal of Neurology* 60, no. 4, 389–443. doi:10:1093/brain/60:4.389.

2. Chalmers, D. J. (1995). Facing up to the problem of consciousness. *Journal of Consciousness Studies* 2, no. 3, 200–219. Retrieved from http://cogprints.org/316/1/consciousness.html.

3. Koch, C., and Tononi, G. (2008). Can machines be conscious? *Spectrum, IEEE* 45, no. 6, 55–59. Retrieved from http://ieeexplore.ieee.org/xpls/abs_all.jsp?arnumber=4531463.

4. However, we do know that neurons can interface very well with silicon systems. The pioneering work of Peter Fromherz, for example, has shown the beauty of a "neurochip" whereby connections are readily made on a circuit board between neurons and silicon nodes. Similarly, if brain cells are able to function in a hybrid device in this way, then the reverse may not be surprising: artificial implants in the brain are already possible and are achieving astonishing effects. For example, Miguel Nicolelis of Duke University has developed a system whereby quadriplegic patients can, through devices implanted in their brain, generate electronic signatures that would normally

precede various movements. These electronic signals are then recognized by a computer that can operate an artificial limb, so that a person paralyzed from the neck down can "will" a movement. However, these "neuronal prostheses" are far from the silicon takeover of the brain envisaged in the thought experiment. While silicon-carbon interfacing is possible, at least for the final execution of a movement, namely brain output stimulating brain muscle, it should not be conflated with the neuron-neuron interactions that underlie cognitive processes, or confused with artificial intelligence.

5. Damasio, A. R., Everitt, B. J., and Bishop, D. (1996). The somatic marker hypothesis and the possible functions of the prefrontal cortex. *Philosophical Transactions of the Royal Society of London. Series B: Biological Sciences* 351, no. 1346, 1413–1420. doi:10:1098/rstb.1996:0125.

6. Turing, A. M. (1950). Computing machinery and intelligence. *Mind* 59, no. 236, 433–460. Retrieved from http://cogprints.org/499/1/turing.html.

7. For example, Rees, G., Kreiman, G., and Koch, C. (2002). Neural correlates of consciousness in humans. *Nature Reviews Neuroscience* 3, no. 4, 261–270. Tononi, G., and Koch, C. (2008). The neural correlates of consciousness. *Annals of the New York Academy of Sciences* 1124, no. 1, 239–261.

8. Koch, C., and Greenfield, S. (2007). How does consciousness happen? *Scientific American* 297, no. 4, 76–83. Retrieved from http://www.sciamdigital .com/index.cfm?fa=ExtServices.GspDownloadIssueView&ARTICLEID_CHAR =E0E902FE-3048-8A5E-1061447DA58B3813.

9. Greenfield, S. A. (2001). *The private life of the brain: Emotions, consciousness, and the secret of the self.* New York: Wiley.

10. James, W. (1890). *The principles of psychology*, Chapter 13. Retrieved from http://psychclassics.yorku.ca/James/Principles/prin13.htm.

11. René Descartes (1596–1650), often dubbed "the father of modern philosophy," argued the case that humans are distinctly different from other animals and the rest of the natural world: our unique mind can be attributed to language and reason, features that set our species apart from the rest of the animal kingdom. Descartes suggested that the demonstrative behaviors of all nonhuman creatures can be explained without having to bother ascribing minds and consciousness to them. He concluded that nonhuman animals can be regarded as no more than machines, with parts assembled in intricate ways. However, although humans might have minds and consciousness, these phenomena would be separate from the mechanistic working of the body: "To explain these functions, then, it is not necessary to conceive of any vegetative or sensitive soul, or any other principle of movement or life, other than its blood and its spirits which are agitated by the heat of the fire that burns continuously in its heart, and which is of the same nature as those fires that occur in inanimate bodies." This notion of a mechanistic physical body extended to the mechanics of the physical brain. For Descartes, the prototypical dualist, it would mean that the physical brain was distinct from the mind and consciousness, which was left largely undefined and unexplored. More recently, in the twentieth century the advent of the computer brought with it the opportunity to jettison the notion of some airy-fairy parallel consciousness, and instead to ascribe everything to mechanistic processes. (Descartes, R. [1994]. The treatise on man. In S. Gaukroger (Ed.), *The world and other writings*, pp. 119–169. Retrieved from http:// www2.dsu.nodak.edu/users/dmeier/31243550-Descartes-The-World-and -Other-Writings.pdf, p. 169.)

12. The definition of intelligence is no mere semantic quibble, but would extend to wider moral questions. For example, Hume was at odds with Kant in assuming that intelligence does not necessarily imply moral values and vice versa. However, surely this dilemma once again depends on how we define intelligence. If we take the simple computational concept of *g,* prowess in IQ tests, then Hume would be correct: after all, why should a simple linear process be predicated on anything other than the rules of the game? But if we take the wider view of intelligence, as I would argue, to imply understanding, then perhaps Kant would be more accurate in viewing intelligence as an understanding that would imply an awareness of the link to particular values.

13. Horn, J. L., and Cattell, R. B. (1967). Age differences in fluid and crystallized intelligence. *Acta Psychologica* 26, 107–129. Retrieved from http://www.sciencedirect.com/science/article/pii/000169186790011X.

14. This idea of true, deep mental abilities arising from neuronal connections would fit with the finding mentioned earlier that gifted children do indeed display greater neuronal connectivity.

15. Greenfield, S. A. (2011). You and me: The neuroscience of identity. London: Notting Hill.

16. Blake, W. (c. 1803). "Auguries of Innocence." Retrieved from http://www.bartleby.com/41/356.html.

CHAPTER 8. OUT OF YOUR MIND

1. In the mid-twentieth century an American physician, Paul MacLean, developed a theory to explain the inexplicable collective behavior of the crowds at the Nuremburg rallies during the Nazi era. MacLean's reasoning was that, anatomically, the brain could be compartmentalized into three evolutionary stages: the *reptilian* brain, consisting of the inner core, the basic part of the brain; layered onto that would be the *mammalian* brain, including areas such as the amygdala and hippocampus; and finally, constituting the most sophisticated level of all, would be the cortex, the outer layer of the brain, which is the monopoly of the *neo-mammalian* species. MacLean argued that these three layers represented increasing degrees of sophistication in mental processes. The reptilian brain underpinned very primitive urges, these being channeled into the appropriate context by virtue of the mammalian brain, while the neo-mammalian brain imposed further refinements, even rules, on how one might behave. This hierarchy of three levels corresponds quite neatly to Freud's notion of the atavistic id, the mediating ego, and the moralistic superego. According to MacLean, emotions were suppressed by logic and reason for most of the time, but within the intermediate limbic system, which he saw as centers for emotions normally held in check by a logical cortex, regions can also play a key role in that most sensible of activities, memory. Conversely, disruptions to the cortex, especially the prefrontal cortex, can be linked to emotional disturbances, such as those seen in addiction, obesity, and schizophrenia. Sadly, however, such simple compartmentalization doesn't stand up to the anatomical and physiological practicalities of what we now know the brain, and indeed the mind, to be capable of. Nonetheless, this theory is useful at a more metaphorical level. According to MacLean, the seemingly blind aggression of the Nuremberg crowds could

therefore be accounted for by the breakdown in the anatomical hierarchy of the "triune brain" (MacLean, P. D. [1985]. Evolutionary psychiatry and the triune brain. *Psychological Medicine* 15, no. 2, 219–221. doi:10:1017/S0033291700023485).

2. Greenfield, S. A. (2008). *I.D.: The quest for meaning in the 21st century.* London: Hodder & Stoughton.

3. Olds, J., and Milner, P. (1954). Positive reinforcement produced by electrical stimulation of septal area and other regions of rat brain. *Journal of Comparative and Physiological Psychology* 47, no. 6, 419–427. Retrieved from http://www.wadsworth.com/psychology_d/templates/student_resources/0155060678_rathus/ps/ps02.html.

4. O'Driscoll, K., and Leach, J. P. (1998). "No longer Gage": An iron bar through the head: Early observations of personality change after injury to the prefrontal cortex. *BMJ* 317, no. 7174, 1673–1674. Retrieved from http://www.ncbi.nlm.nih.gov/pmc/articles/PMC1114479/#__ffn_sectitle.

5. O'Driscoll and Leach, 1998, p. 1673.

6. Tsujimoto, S. (2008). The prefrontal cortex: functional neural development during early childhood. *The Neuroscientist* 14, no. 4, 345–358. doi:10:1177/107385840831600.

7. Sturman, D. A., and Moghaddam, B. (2011). The neurobiology of adolescence: changes in brain architecture, functional dynamics, and behavioral tendencies. *Neuroscience & Biobehavioral Reviews* 35, no. 8, 1704–1712. doi:10:1016/j.neubiorev.2011:04.003.

8. Steinberg, L. (2008). A social neuroscience perspective on adolescent risk-taking. *Developmental Review* 28, no. 1, 78–106. doi: 10:1016/j.dr.2007:08.002.

9. Casey, B. J., Getz, S., and Galvan, A. (2008). The adolescent brain. *Developmental Review* 28, no. 1, 62–77. doi: 10:1016/j.dr.2007:08.003.

10. Casey, Getz, and Galvan, 2008.

11. Callicott, J. H., Bertolino, A., Mattay, V. S., Langheim, F. J., Duyn, J., Coppola, R., . . . and Weinberger, D. R. (2000). Physiological dysfunction of the dorsolateral prefrontal cortex in schizophrenia revisited. *Cerebral Cortex* 10, no. 11, 1078–1092. doi:10:1093/cercor/10:11.1078.

12. Volkow, N. D., Wang, G. J., Telang, F., Fowler, J. S., Goldstein, R. Z., Alia-Klein, N., . . . and Pradhan, K. (2008). Inverse association between BMI and prefrontal metabolic activity in healthy adults. *Obesity* 17, no. 1, 60–65. doi:10:1038/oby.2008:469.

13. Pignatti, R., Bertella, L., Albani, G., Mauro, A., Molinari, E., and Semenza, C. (2006). Decision-making in obesity: A study using the Gambling Task. *Eating and Weight Disorders: EWD* 11, no. 3, 126. Retrieved from http://www.ncbi.nlm.nih.gov/pubmed/17075239.

14. Dang-Vu, T. T., Schabus, M., Desseilles, M., Sterpenich, V., Bonjean, M., and Maquet, P. (2010). Functional neuroimaging insights into the physiology of human sleep. *Sleep* 33, no. 12, 1589–1603. Retrieved from http://www.ncbi.nlm.nih.gov/pmc/articles/PMC2982729/#__ffn_sectitle.

15. Greenfield, S. (2011). *You and me: The Neuroscience of identity.* London: Notting Hill.

CHAPTER 9. THE *SOMETHING* ABOUT SOCIAL NETWORKING

1. O'Connell, R. (May 12, 2011). The pros and cons of deleting your Facebook [blog post]. Retrieved from http://thoughtcatalog.com/2011/the-pros-and -cons-to-deleting-your-facebook.
2. Hampton, K. N., Goulet, L. S., Rainie, L., and Purcell, K. (2011). *Social networking sites and our lives.* Retrieved from http://pewinternet.org/Reports/ 2011/Technology-and-social-networks.aspx.
3. Badoo. (April 25, 2012). Generation lonely? 39 percent of Americans spend more time socializing online than face-to-face. Retrieved from http://corp .badoo.com/he/entry/press/54.
4. Quoted in McCullagh, D. (March 12, 2010). Why no one cares about privacy anymore. Retrieved from http://www.cnet.com/uk/news/why-no-one-cares -about-privacy-anymore.
5. Protalinski, E. (May 1, 2013) Facebook passes 1.11 billion monthly active users, 751 million mobile users, and 665 million daily users. Retrieved from http://thenextweb.com/facebook/2013/05/01/facebook-passes-1-11-billion -monthly-active-users-751-million-mobile-users-and-665-million-daily -users.
6. Anderson, B., Fagan, P., Woodnutt, T., and Chamorro-Prezumic, T. (2012). Facebook psychology: Popular questions answered by research. *Psychology of Popular Media Culture* 1, no. 1, 23–37. doi:10:1037/a0026452.
7. Manago, A. M., Taylor, T., and Greenfield, P. M. (2012). Me and my 400 friends: The anatomy of college students' Facebook networks, their communication patterns, and well-being. *Developmental Psychology* 48, no. 2, 369–380. doi: 10:1037/a0026338.
8. Grieve, R., Indian, M., Witteveen, K., Tolan, G. A., and Marrington, J. (2013). Face-to-face or Facebook: Can social connectedness be derived online? *Computers in Human Behavior* 29, no. 3, 604–609. doi:10:1016/j.chb.2012:11.017.
9. Quoted in Cohen, J. (February 1, 2012). Facebook officially files SEC documents for $5B offer [blog post]. Retrieved from http://allfacebook.com/ facebook-files-ipo_b76165.
10. Teilhard de Chardin, P. (1964). *The future of man.* London: Collins.
11. Rutledge, T., et al. (2008). Social networks and incident stroke among women with suspected myocardial ischemia. *Psychosomatic Medicine* 70, no. 3, 282–287. doi:10:1097/PSY.0b013e3181656e09.
12. Cole, S. W., Hawkley, L. C., Arevalo, J. M. G., and Cacioppo, J. T. (2011). Transcript origin analysis identifies antigen-presenting cells as primary targets of socially regulated gene expression in leukocytes. *PNAS* 108, no. 7, 3080–3085. doi:10:1073/pnas.1014218108.
13. Norman, G. J., Cacioppo, J. T., Morris, J. S., Malarkey, W. B., Berntson, G. G., and DeVries, A. C. (2011). Oxytocin increases autonomic cardiac control: Moderation by loneliness. *Biological Psychology* 86, no. 3, 174–180.
14. Klinenberg, E. (March 30, 2012). I want to be alone: The rise of solo living. Retrieved from http://www.theguardian.com/lifeandstyle/2012/mar/30/the -rise-of-solo-living.
15. Sigman, A. (2009). Well connected? The biological implications of "social networking." *Biologist* 56, no. 1, 14–20. Retrieved from http://www .aricsigman.com/IMAGES/Sigman_lo.pdf.
16. Penenberg, A. L. (July 1, 2010). Social networking affects brains like falling in

love [blog post]. Retrieved from http://www.fastcompany.com/1659062/ social-networking-affects-brains-falling-loved.

17. Wilson, R. E., Gosling, S. D., and Graham, L. T. (2012). A review of Facebook research in the social sciences. *Perspectives on Psychological Science* 7, no. 3, 203–220. doi:10:1177/1745691612442904.

18. Burke, M., Marlow, C., and Lento, T. (2010). Social network activity and social well-being. *Proceedings of the SIGCHI Conference on Human Factors in Computing System*, 1909–1912. doi:10:1145/1753326:1753613.

19. Clayton, R. B., Osborne, R. E., Miller, B. K., and Oberle, C. D. (2013). Loneliness, anxiousness, and substance use as predictors of Facebook use. *Computers in Human Behavior* 29, no. 3, 687–693. doi:10:1016/j.chb.2012:12.002.

20. Skues, J. L., Williams, B., and Wise, L. (2012). The effects of personality traits, self-esteem, loneliness, and narcissism on Facebook use amongst university students. *Computers in Human Behavior* 28, no. 6, 2414–2419. doi:10:1016/ j.chb.2012:07.012.

21. Watson, R. (2010). *Future files: A brief history of the next 50 years.* London: Nicholas Brealey.

22. Anderson et al., 2012.

23. Oldmeadow, J. A., Quinn, S., and Kowert, R. (2013). Attachment style, social skills, and Facebook use amongst adults. *Computers in Human Behavior* 29, no. 3, 1142–1149. doi:10:1016/j.chb.2012:10.006.

24. Bowlby, J. (1969). *Attachment and loss,* Vol. 1: *Loss.* New York: Basic Books, p.194.

25. Oldmeadow, Quinn, and Kowert, 2013.

26. Skues, Williams, and Wise, 2012.

27. Tamir, D. I., and Mitchell, J. P. (2012). Disclosing information about the self is intrinsically rewarding. *PNAS* 109, no. 21, 8038–8043. doi:10:1073/pnas .1202129109.

28. Arbitron and Edison Research. (April 2013). *The infinite dial 2013: Navigating digital platforms.* Retrieved from http://www.edisonresearch.com/wp -content/uploads/2013/04/Edison_Research_Arbitron_Infinite_Dial_2013.pdf.

29. Jiang, L. C., Bazarova, N. N., and Hancock, J. T. (2011). The disclosure-intimacy link in computer-mediated communication: An attributional extension of the hyperpersonal model. *Human Communication Research* 37, no. 1, 58–77. doi:10:1111/j.1468-2958:2010:01393.x; boyd, d. m., and Ellison, N. B. (2007). Social networking sites: Definition, history, and scholarship. *Journal of Computer-Mediated Communication* 13, no. 1, 210–230. doi: 10:1111/j.1083-6101:2007:00393.x.

30. Trepte, S., and Reinecke, L. (2013). The reciprocal effects of social network site use and the disposition for self-disclosure: A longitudinal study. *Computers in Human Behavior* 29, no. 3, 1102–1112. doi:10:1016/j.chb.2012:10.002.

31. Tamir and Mitchell, 2012.

32. Mauri, M., Cipresso, P., Balgera, A., Villamira, M., and Riva, G. (2011). Why is Facebook so successful? Psychophysiological measures describe a core flow state while using Facebook. *Cyberpsychology, Behavior, and Social Networking* 14, no. 12, 723–731. doi:10:1089/cyber.2010:0377, p. 1.

33. Weinschenk, S. (November 7, 2009). 100 things you should know about people. #8 Dopamine makes you addicted to seeking information [blog post]. Retrieved from http://www.blog.theteamw.com/2009/11/07/100-things-you -should-know-about-people-8-dopamine-makes-us-addicted-to-seeking -information.

34. O'Doherty, J., Deichmann, R., Critchley, H., and Dolan, R. J. (2002). Neural responses during anticipation of a primary taste reward. *Neuron* 33, no. 5, 815–826. doi:10:1016/S0896-6273(02)00603-7.
35. Does Facebook "addiction" really exist? American psychologist Michael Fenichel has suggested that, like gambling or alcohol, Facebook might have its very own version of addiction. He describes the all too familiar situation in which Facebook usage can trump daily activities such as waking up, getting dressed, using the telephone, or checking email. Fenichel has accordingly introduced a new term to describe such a state: Facebook addiction disorder, or FAD. He defines FAD as a condition where hours are spent on Facebook, to the extent that the healthy balance of the individual's life is affected. Fenichel claims that approximately 350 million people are suffering from the condition, which can be detected through a simple set of six criteria. People who are victims of the disorder must have at least two or three of the following criteria during a six- to eight-month period. For the family members and friends who think they are dealing with an addict, a sign to look out for is, apparently, multiple Facebook windows open. Three or more windows, bizarrely, confirms that they are indeed suffering from this condition. There remains no empirical evidence that Facebook addiction disorder exists (Fenichel, M. [n.d.]. Facebook addiction disorder [FAD]. Retrieved from http://www.fenichel.com/facebook).
36. Johnson, D. E., Guthrie, D., Smyke, A. T., Koga, S. F., Fox, N. A., Zeanah, C. H., and Nelson, C. A. (2010). Growth and associations between auxology, caregiving environment, and cognition in socially deprived Romanian children randomized to foster vs. ongoing institutional care. *Archives of Paediatrics & Adolescent Medicine* 164, no. 6, 507–516. doi:10:1001/archpediatrics.2010:56.
37. Oldmeadow, Quinn, and Kowert, 2013.
38. Dumon, M. (October 18, 2011). Meet George Clooney's new girl: Stacy Keibler. Retrieved from http://www.examiner.com/article/meet-george -clooney-s-new-girl-stacy-keibler.
39. Harkaway, N. (2012). *The blind giant: Being human in a digital world*. London: Vintage.
40. McCullagh, 2010.
41. McAfee. (2010). *The secret online lives of teens*. Retrieved from http://us.mcafee .com/en-us/local/docs/lives_of_teens.pdf.
42. Arbitron and Edison Research. (April 2013). *The infinite dial 2013: Navigating digital platforms*. Retrieved from http://www.edisonresearch.com/wp-content/ uploads/2013/04/Edison_Research_Arbitron_Infinite_Dial_2013.pdf.

CHAPTER 10. SOCIAL NETWORKING AND IDENTITY

1. Government Office for Science. (2013). *Future identities: Changing identities in the UK: The next 10 years*. Retrieved from https://www.gov.uk/government/ uploads/system/uploads/attachment_data/file/273966/13-523-future-identities -changing-identities-report.pdf.
2. Amichai-Hamburger, Y., Wainapel, G., and Fox, S. (2002). "On the Internet no one knows I'm an introvert": Extroversion, neuroticism, and Internet interaction. *CyberPsychology & Behavior* 5, no. 2, 125–128. Retrieved from http://www.ncbi.nlm.nih.gov/pubmed/12025878.
3. Suler, J. (2004). The online disinhibition effect. *CyberPsychology & Behavior* 7,

no. 3, 321–326. doi:10:1089/1094931041291295. Christopherson, K. M. (2007). The positive and negative implications of anonymity in Internet social interactions: "On the Internet, nobody knows you're a dog." *Computers in Human Behavior* 23, no. 6, 3038–3056. doi:10:1016/j.chb.2006:09.001.

4. Zhao, S., Grasmuck, S., and Martin, J. (2008). Identity construction on Facebook: Digital empowerment in anchored relationships. *Computers in Human Behavior* 24, 1816–1836. doi:10:1016/j.chb.2008:02.012.

5. boyd, d. m., and Ellison, N. B. (2007). Social networking sites: Definition, history, and scholarship. *Journal of Computer-Mediated Communication* 13, no. 1, 210–230. doi:10:1111/j.1083–6101:2007:00393.x.

6. boyd, d. m., and Ellison, N. B., 2007, p. 211.

7. What names are allowed on Facebook? (n.d.). Retrieved from https://www.facebook.com/help/112146705538576?q=name&sid=09QL15Kz6090K35pZ.

8. Rogers, C. (1951). *Client-centered therapy*. Boston: Houghton-Mifflin.

9. Bargh, J. A., McKenna, K. Y. A., and Fitzsimons, G. M. (2002). Can you see the real me? Activation and expression of the "true self" on the Internet. *Journal of Social Issues* 58, no. 1, 33–48. doi:10:1111/1540–4560:00247.

10. McKenna, K. Y. A., Green, A. S., and Gleason, M. E. J. (2002). Relationship formation on the Internet: What's the big attraction? *Journal of Social Issues* 58, no. 1, 9–31. doi:10:1111/1540–4560:00246.

11. McKenna, Green, and Gleason, 2002.

12. Tosun, L. P. (2012). Motives for Facebook use and expressing "true self" on the Internet. *Computers in Human Behavior* 28, 1510–1517. doi:10:1016/j.chb.2012:03.018.

13. Tosun, L. P., and Lajunen, T. (2009). Why do young adults develop a passion for Internet activities? The associations among personality, revealing "true self" on the Internet, and passion for the Internet. *CyberPsychology & Behavior* 12, no. 4, 401–406. doi:10:1089/cpb.2009:0006.

14. Zhao, Grasmuck, and Martin, 2008.

15. Siibak, A. (2009). Constructing the self through the photo selection: Visual impression management on social networking websites. *Cyberpsychology: Journal of Psychosocial Research on Cyberspace* 3, no. 1, article 1. Retrieved from: http://cyberpsychology.eu/view.php?cisloclanku=2009061501&article=1.

16. Goffman, E. (1959). *The presentation of self in everyday life*. New York: Overlook.

17. d. m. boyd interviewed in Rosen, L. D. (2012). *iDisorder: Understanding our obsession with technology and overcoming its hold on us*. Harmondsworth, UK: Palgrave Macmillan, p. 34.

18. Zhao, Grasmuck, and Martin, 2008.

19. While the majority of social networking research has focused on identity specifically on Facebook, there has been a proposal that, as different social networking platforms offer different types of social networking to users, so different identities might be managed on different social networking sites. For example, LinkedIn may be used to develop the hoped-for professional self, whereas Facebook is the platform to display the hoped-for social self (van Dijck, J. [2013]. "You have one identity": Performing the self on Facebook and LinkedIn. *Media, Culture & Society* 35, no. 2, 199–215. doi:10:1177/0163443712468605).

20. Zhao, Grasmuck, and Martin, 2008.

21. Back, M. D., Stopfer, J. M., Vazire, S., Gaddis, S., Schmukle, S. C., Egloff, B., and Gosling, S. D. (2010). Facebook profiles reflect actual personality, not

self-idealization. *Psychological Science* 21, no. 3, 372–374. doi: 10:1177/0956797609360756.

22. Rosen, 2012.
23. Buffardi, L. E., and Campbell, W. K. (2008). Narcissism and social networking Web sites. *Personality and Social Psychology Bulletin* 34, 1303–1314. doi: 10:1177/0146167208320061. Mehdizadeh, S. (2010). Self-Presentation 2.0: Narcissism and self-esteem on Facebook. *Cyberpsychology, Behavior, and Social Networking* 13, no. 4, 357–364. doi:10:1089/cyber.2009:0257. Ryan, T., and Xenos, S. (2011). Who uses Facebook? An investigation into the relationship between the Big Five, shyness, narcissism, loneliness, and Facebook usage. *Computers in Human Behavior* 27, 1658–1664. doi:10:1016/j.chb.2011:02.004. Twenge, J. M., Konrath, S., Foster, J. D., Campbell, W. K., and Bushman, B. J. (2008). Egos inflating over time: A cross-temporal meta-analysis of the narcissistic personality inventory. *Journal of Personality* 76, no. 4, 875–902. doi:10:1111/j.1467–6494:2008:00507.x.
24. Twenge et al., 2008.
25. In a study of Twitter users by Mor Naaman and his team from Rutgers, the subjects fell into two categories: "meformers" and "informers." As their name suggests, the meformers posted endless updates on their own thoughts and feelings, while informers lived up to their particular name by sharing information and interacting more with followers. Of those studied, 80 percent of the subjects were classified as meformers, fitting well into the profile of our current narcissistic era (Naaman, M., Boase, J., and Lai, C. H. [2010]. Is it really about me? Message content in social awareness streams. *Proceedings of the 2010 ACM Conference on Computer Supported Cooperative Work*, 189–192. doi:10:1145/1718918:1718953).
26. Buffardi and Campbell, 2008. Mehdizadeh, 2010. Ryan and Xenos, 2011. Naaman, Boase, and Lai, 2010. McKinney, B. C., Kelly, L., and Duran, R. L. (2012). Narcissism or openness? College students' use of Facebook and Twitter. *Communication Research Reports* 29, no. 2, 108–118. doi:10:1080/08824096:2012:666919. Bergman, M., Fearrington, M. E., Davenport, S. W., and Bergman, J. Z. (2011). Millennials, narcissism, and social networking: What narcissists do on social networking sites and why. *Personality and Individual Differences* 50, 706–711. doi:10:1016/j.paid.2010:12.022. Carpenter, C. J. (2012). Narcissism on Facebook: Self-promotional and anti-social behavior. *Personality and Individual Differences* 52, no. 4, 482–486. doi:10:1016/j.paid.2011:11.011. Panek, E. T., Nardis, Y., and Konrath, S. (2013). Defining social networking sites and measuring their use: How narcissists differ in their use of Facebook and Twitter. *Computers in Human Behavior* 29, no. 5, 2004–2012. doi:10:1016/j.chb.2013:04.012.
27. Raskin, R., and Terry, H. (1988). A principal-components analysis of the Narcissistic Personality Inventory and further evidence of its construct validity. *Journal of Personality and Social Psychology* 54, no. 5, 890–902. doi:10:1037/00 22–3514:54.5:890.
28. Panek, Nardis, and Konrath, 2013.
29. A potential upside from regularly viewing and modifying your social networking identity may be increased self-esteem. However, earlier research has shown that inducing self-awareness via a mirror can induce a negative mood, particularly for women (Fejfar, M. C., and Hoyle, R. H. [2000]. Effect of private self-awareness on negative affect and self-referent attribution: A quantitative review. *Personality and Social Psychology Review* 4, no. 2, 132–

142. doi:10:1207/S15327957PSPR0402_02), so, for some, viewing one's social networking profile may be the equivalent of an online mirror, and may have negative effects on self-esteem. But Facebook is not a true mirror, displaying an unedited image of ourselves; it is a modified and managed mirror that reflects back our self-edited best version of ourselves, and thus has the potential to be a rose-tinted distortion. Subsequently, research has found that viewing one's own Facebook profile results in higher self-esteem levels compared to those who looked in a mirror, with those who edited their profiles during a short testing period possessing the highest levels of self-esteem (Tazghini, S., and Siedlecki, K. L. [2013]. A mixed method approach to examining Facebook use and its relationship to self-esteem. *Computers in Human Behavior* 29, no. 3, 827–832. doi:10:1016/j.chb.2012:11.010). Unsurprisingly, it seems that the ability to create and present the ideal version of yourself has positive effects on self-esteem. While older forms of media such as glossy fashion magazines and TV shows increase body image issues, particularly in women, research has shown that the strongest media-related predictor of these issues is now social network site use (Tiggemann, M., and Miller, J. [2010]. The Internet and adolescent girls' weight satisfaction and drive for thinness. *Sex Roles* 63, 79–90. doi:10:1007/s11199–010–9789-z). Girls who spent more time on Facebook and MySpace displayed higher scores of "drive for thinness," a subscale of an eating disorder diagnostic tool. Facebook use was also linked to girls being less satisfied with their current weight and having higher levels of an internalized thin ideal. These associations were stronger for social networking sites than for traditional culprits of body image disorder in women, such as magazines and TV. Most of the research has been equivocal as to whether social networking can actually promote healthy types of self-esteem. At Canada's York University, psychology student Soraya Mehdizadeh (2010) examined the online habits and personalities of Facebook users at the university, ranging from eighteen to twenty-five years old. Mehdizadeh explored how narcissism and self-esteem related to the various self-promotional contents of a Facebook profile, and found that individuals higher in narcissism and lower in self-esteem spent more time on the site and filled their pages with more self-promotional content. So filling a Facebook page with positive versions of oneself does not appear to do much for an individual's self-esteem. Perhaps it's because, for real reassurance, we all need real-world feedback, the literal and metaphorical pat on the back that comes with voice tone, eye contact, body language, and physical contact. A crucial factor may be the type of online activity involved. One study examined the relationship between self-esteem and Facebook use in a sample of some two hundred college students (Manago, Taylor, and Greenfield, [2012]. Me and my 400 friends: The anatomy of college students' Facebook networks, their communication patterns, and well-being. *Developmental Psychology* 48, no. 2, 369–380. doi:10:1037/a0026338). Results indicated that self-esteem level was related to engaging in different online behaviors. For example, lower self-esteem was associated with feelings of connectedness to Facebook (that is, to the site itself), more frequently untagging oneself in photos, and accepting friend requests from acquaintances or strangers. In contrast, individuals with higher self-esteem were more likely to report that a positive aspect of Facebook was the ability to share pictures, thoughts, and ideas, and that other people's posts could become annoying or bothersome. The conclusion was that individuals with low self-esteem use Facebook to

accrue more friends and manage their profiles. Then again, perhaps large audiences inflate self-esteem, and if so, those who use Facebook to accrue large networks are potentially at risk of developing unhealthy estimates of their own worth (Gonzales, A. L., and Hancock, J. T. [2011]. Mirror, mirror on my Facebook wall: Effects of exposure to Facebook on self-esteem. *Cyberpsychology, Behavior, and Social Networking* 14, nos. 1–2, 79–83. doi: 10:1089/ cyber.2009:0411). Meanwhile, participants who also view others' profile pages do not have as high self-esteem as those who focus solely on their own profiles (Gonzales and Hancock, 2011). Accordingly, another study found that those who focused on their own Facebook page had higher levels of self-esteem than a control group (Gentile, B., Twenge, J. M., Freeman, E. C., and Campbell, W. K. [2012]. The effect of social networking websites on positive self-views: An experimental investigation. *Computers in Human Behavior* 28, no. 5, 1929–1933. doi:10:1016/j.chb.2012:05.012). Again, perhaps predictably, individualistic, self-focused social networking has the strongest association with rating yourself highly. However, these studies may only be highlighting correlations between those who most enjoy Facebook and high self-esteem. Does social networking simply reinforce the high opinion of individuals who already have robust self-esteem, or can it actually increase the levels of self-esteem in those who are not so confident in themselves? A critical determinant of any positive effects of social network use is the type of feedback users receive from their Facebook audience (Valkenburg, P. M., Peter, J., and Schouten, M. A. [2006]. Friend networking sites and their relationship to adolescents' well-being and social self-esteem. *CyberPsychology & Behavior* 9, no. 5, 584–590. doi:10:1089/cpb.2006:9.584).

30. Valkenburg, Peter, and Schouten, 2006.
31. Valkenburg, Peter, and Schouten, 2006.
32. Facebook cull: Top reasons to unfriend someone. (July 3, 2013). Retrieved from http://www.huffingtonpost.co.uk/2013/07/03/facebook-reasons-to -unfriend-someone_n_3541249.html.
33. Forest, A. L., and Wood, J. V. (2012). When social networking is not working: Individuals with low self-esteem recognize but do not reap the benefits of self-disclosure on Facebook. *Psychological Science* 23, no. 3, 295–302. doi:10:1177/0956797611429709.
34. Manago, A. M., Taylor, T., and Greenfield, P. M. (2012). Me and my 400 friends: The anatomy of college students' Facebook networks, their communication patterns, and well-being. *Developmental Psychology* 48, no. 2, 369–380. doi:10:1037/a0026338.
35. Manago, Taylor, and Greenfield, 2012.
36. Qiu, L., Lin, H., Leung, A. K., and Tov, W. (2012). Putting their best foot forward: Emotional disclosure on Facebook. *Cyberpsychology, Behavior, and Social Networking* 15, no. 10, 569–572. doi:10:1089/cyber.2012:0200.
37. Sigman, A. (2009). Well connected? The biological implications of "social networking." *Biologist* 56, no. 1, 14–20. Retrieved from http://www .aricsigman.com/IMAGES/Sigman_lo.pdf.
38. KidScape. (2011). *Young people's cyber life survey*. Retrieved from http://www .kidscape.org.uk/resources/surveys.
39. Kanai, R., Bahrami, B., Roylance, R., and Rees, G. (2011). Online social network size is reflected in human brain structure. *Proceedings of the Royal Society Biological Sciences* 279, no. 1732, 1327–1334. doi:10:1098/rspb.2011: 1959.

40. Turkle, S. (2012). *Alone together: Why we expect more from technology and less from each other.* New York: Basic Books.
41. James, O. (2008). *Affluenza.* London: Vermilion.
42. Krasnova, H., Wenninger, H., Widjaja, T., and Buxmann, P. (2013). Envy on Facebook: A hidden threat to users' life satisfaction? In *11th International Conference on Wirtschaftsinformatik.* Leipzig, Germany.
43. Marshall, T. C. (2012). Facebook surveillance of former romantic partners: Associations with post-breakup recovery and personal growth. *Cyberpsychology, Behavior, and Social Networking* 15, no. 10, 521–526. doi:10:1089/cyber.2012:0125.
44. Tong, S. T., Van Der Heide, B., Langwell, L., and Walther, J. B. (2008). Too much of a good thing? The relationship between number of friends and interpersonal impressions on Facebook. *Journal of Computer-Mediated Communication* 13, 531–549. doi:10:1111/j.1083–6101:2008:00409.x.
45. Klout scores are then supplemented with three nominally more specific measures, which Klout calls "true reach," "amplification," and "network impact." True reach is based on the amount of influence, determined by the number of followers and friends who actively listen and react to one's online messages; the amplification score relates to the likelihood that one's messages will generate actions (retweets, @messages, likes, and comments); the network score reflects the influence that a person's engaged audience has.
46. Llenas, B. (November 3, 2011). Klout CEO Fernandez responds to critics, gives insider tips and thinks ahead. Retrieved from http://latino.foxnews.com/latino/community/2011/11/03/klout-ceo-fernandez-responds-to-critics-gives-tips-and-talks-future.
47. Bates, D. (June 17, 2011). "Leaving Facebook? You can try . . . but 'evil genius' social network won't make it easy." Retrieved from http://www.dailymail.co.uk/sciencetech/article-2004610/Leaving-Facebook-You-try—evil-genius-social-network-wont-make-easy.html.
48. Stieger, S., Burger, C., Bohn, M., and Voracek, M. (in press). Who commits virtual identity suicide? Differences in privacy concerns, Internet addiction, and personality between Facebook users and quitters. *Cyberpsychology, Behavior, and Social Networking.* doi:10:1089/cyber.2012:0323.

CHAPTER 11. SOCIAL NETWORKING AND RELATIONSHIPS

1. Plato. (1925). *Plato in Twelve Volumes.* (H. N. Fowler, Trans.). Cambridge, MA: Harvard University Press; London: William Heinemann Ltd. Retrieved from http://www.perseus.tufts.edu/hopper/text?doc=Perseus%3Atext%3A1999:01.0174%3Atext%3DPhaedrus%3Apage%3D275.
2. Quoted in http://stakeholders.ofcom.org.uk/market-data-research/market-data/communications-market-reports/cmr12/market-context.
3. Ofcom. (July 18, 2012). *The communications market report 2012.* Retrieved from http://media.ofcom.org.uk/files/2012/07/CMR_analyst_briefing_180712.pdf.
4. Ofcom, 2012.
5. Seltzer, L. J., Prososki, A. R., Ziegler, T. E., and Pollak, S. D. (2012). Instant messages vs. speech: Hormones and why we still need to hear each other. *Evolution and Human Behavior* 33, 42–45. doi:10:1016/j.evolhumbehav.2011:05.004.

6. Lord, L. (January 14, 2013). Generation Y, dating and technology: Digital natives struggle to connect offline. Retrieved from http://www.huffingtonpost.ca/2013/01/14/generation-y-online-dating-technology-relationships_n_2457722.html.

7. Turkle, S. (2012). *Alone together: Why we expect more from technology and less from each other.* New York: Basic Books, p. 1.

8. Howard-Jones, P. (2011). *The impact of digital technologies on human wellbeing: Evidence from the sciences of mind and brain.* Retrieved from http://www.nominettrust.org.uk/sites/default/files/NT%20SoA%20-%20The%20impact%20of%20digital%20technologies%20on%20human%20wellbeing.pdf, p. 17.

9. Burke, M., Kraut, R., and Marlow, C. (2011). Social capital on Facebook: Differentiating uses and users. *Proceedings of the SIGCHI Conference on Human Factors in Computing Systems,* 573–580. doi:10:1145/1978942:1979023.

10. Bessière, K., Kiesler, S., Kraut, R., and Boneva, B. S. (2008). Effects of Internet use and social resources on changes in depression. *Information, Communication, and Society* 11, no. 1, 47–70. doi:10:1080/13691180701858851.

11. Valkenburg, P. M., and Peter, J. (2007). Preadolescents' and adolescents' online communication and their closeness to friends. *Developmental Psychology* 43, no. 2, 267–277. doi:10:1037/0012–1649:43.2:267.

12. Grieve, R., Indian, M., Witteveen, K., Tolan, G. A., and Marrington, J. (2013). Face-to-face or Facebook: Can social connectedness be derived online? *Computers in Human Behavior* 29, no. 3, 604–609. doi:10:1016/j.chb.2012:11.017.

13. Pollet, T. V., Roberts, S. G. B., and Dunbar, R. I. M. (2011). Use of social network sites and instant messaging does not lead to increased offline social network size, or to emotionally closer relationships with offline network members. *Cyberpsychology, Behavior, and Social Networking* 14, no. 4, 253–258. doi:10:1089/cyber.2010:0161.

14. However, like humans, chimpanzees have "mirror neurons" at work in their brains, cells such as the ones related to eating, which can be activated simply by observing another chimp eating. It seems that the observer chimp has "empathized" with the luckier counterpart actually enjoying a real grape. So the ability to empathize is quite a basic component of the tool kit of the primate brain. Di Pellegrino, G., Fadiga, L., Fogassi, L., Gallese, V., and Rizzolatti, G. (1992). Understanding motor events: A neurophysiological study. *Experimental Brain Research* 91, 176–180. Retrieved from http://www.fulminiesaette.it/_uploads/foto/legame/DiPellegrinoEBR92.pdf.

15. Sagi, A., and Hoffman, M. L. (1976). Empathic distress in the newborn. *Developmental Psychology* 12, no. 2, 175–176. doi:10:1037/0012–1649:12.2:175.

16. Knafo, A., Zahn-Waxler, C., Van Hulle, C., Robinson, J. L., and Rhee, S. H. (2008). The developmental origins of a disposition toward empathy: Genetic and environmental contributions. *Emotion* 8, no. 6, 737–752. doi:10:1037/a001417.

17. Ioannidou, F., and Konstantikaki, V. (2008). Empathy and emotional intelligence: What is it really about? *International Journal of Caring Science* 1, no. 3, 118–123. Retrieved from http://www.caringsciences.org/volume001/issue3/Vol1_Issue3_03_Ioannidou_Abstract.pdf, p. 118.

18. Konrath, S. H., O'Brien, E. H., and Hsing, C. (2011). Changes in dispositional empathy in American college students over time: A meta-analysis. *Personality and Social Psychology Review* 15, no. 2, 180–198. doi:10:1177/1088868310377395.

19. McPherson, M., Smith-Lovin, L., and Brashears, M. E. (2006). Social isolation in America: Changes in core discussion networks over two decades. *American Sociological Review* 71, no. 3, 353–375. doi:10:1177/000312240607100301.

20. Rosen, L. D. (2012). *iDisorder: Understanding our obsession with technology and overcoming its hold on us.* New York: Palgrave Macmillan.

21. Engelberg, E., and Sjöberg, L. (2004). Internet use, social skills, and adjustment. *CyberPsychology & Behavior* 7, no. 1, 41–47. doi:10:1089/10949310 4322820101.

22. He, J. B., Liu, C. J., Guo, Y. Y., and Zhao, L. (2011). Deficits in early stage face perception in excessive Internet users. *Cyberpsychology, Behavior, and Social Networking* 14, no. 5, 303–308. doi:10:1089/cyber.2009:0333.

23. McDowell, M. J. (2004). Autism, early narcissistic injury and self-organization: A role for the image of the mother's eyes? *Journal of Analytical Psychology* 49, no. 4, 495–520. doi:10:1111/j.0021–8774:2004:00481.x.

24. Waldman, M., Nicholson, S., and Adilov, N. (2006). *Does television cause autism?* Working Paper No. 12632. Cambridge, MA: National Bureau of Economic Research. Waldman, M., Nicholson, S., and Adilov, N. (2012). *Positive and negative mental health consequences of early childhood television watching.* Working Paper No. 17786. Cambridge, MA: National Bureau of Economic Research.

25. Hertz-Picciotto, I., and Delwiche, L. (2009). The rise in autism and the role of age at diagnosis. *Epidemiology* 20, no. 1, 84–90. doi:10:1097/EDE.0b013e 3181902d15.

26. Amodio, D. M., and Frith, C. D. (2006). Meeting of minds: The medial frontal cortex and social cognition. *Nature Reviews Neuroscience* 7, 268–277. doi:10:1038/nrn1884.

27. Finkenauer, C., Pollman, M. M. H., Begeer, S., and Kerkhof, P. (2012). Examining the link between autistic traits and compulsive Internet use in a nonclinical sample. *Journal of Autism and Developmental Disorders* 42, 2252–2256. doi:10:1007/s10803-012-1465-4.

28. About ECHOES. (n.d.). Retrieved from http://echoes2.org/?q=node/2.

29. Clayton, R. B., Nagurney, A., and Smith, J. R. (2013). Cheating, breakup, and divorce: Is Facebook use to blame? *Cyberpsychology, Behavior, and Social Networking* 16, no. 10, 717–720. doi:10:1089/cyber.2012:0424.

30. Anderson, B., Fagan, P., Woodnutt, T., and Chamorro-Prezumic, T. (2012). Facebook psychology: Popular questions answered by research. *Psychology of Popular Media Culture* 1, no. 1, 23–37. doi:10:1037/a0026452.

31. Marshall, T. C. (2012). Facebook surveillance of former romantic partners: Associations with post-breakup recovery and personal growth. *Cyberpsychology, Behavior, and Social Networking* 15, no. 10, 521–526. doi:10:1089/cyber .2012:0125.

32. Marshall, 2012, p. 521.

33. Tokunaga, R. S. (2011). Social networking site or social surveillance site? Understanding the use of interpersonal electronic surveillance in romantic relationships. *Computers in Human Behavior* 27, no. 2, 705–713. doi:10:1016/ j.chb.2010:08.014.

34. Stern, L. A., and Taylor, K. (2007). Social networking on Facebook. *Journal of the Communication, Speech & Theatre Association of North Dakota* 20, 9–20. Retrieved from http://www.cstand.org/userfiles/file/journal/2007.pdf#page=9.

35. Muise, A., Christofides, E., and Desmarais, S. (2009). More information than you ever wanted: Does Facebook bring out the green-eyed monster of jeal-

ousy? *CyberPsychology & Behavior* 12, no. 4, 441–444. doi:10:1089/cpb.2008:0263. Muscanell, N. L., Guadagno, R. E., Rice, L., and Murphy S. (2013). Don't it make my brown eyes green? An analysis of Facebook use and romantic jealousy. *Cyberpsychology, Behavior and Social Networking* 16, no. 4, 237–242. doi:10:1089/cyber.2012:0411.

36. Facebook fuelling divorce, research claims. (December 21, 2009). Retrieved from http://www.telegraph.co.uk/technology/facebook/6857918/Facebook-fuelling-divorce-research-claims.html.

37. Facebook fuelling divorce, research claims, 2009.

CHAPTER 12. SOCIAL NETWORKING AND SOCIETY

1. Maag, C. (November 2012). A hoax turned fatal draws anger but no charges. Retrieved from: http://www.nytimes.com/2007/11/28/us/28hoax.html?_r=0.

2. LeBlanc, J. C. (October 2012). Cyberbullying and suicide: A retrospective analysis of 22 cases. *AAP Experience National Conference & Exhibition Council on School Health*. Retrieved from https://aap.confex.com/aap/2012/webprogram/Paper18782.html.

3. Tokunaga, R. S. (2010). Following you home from school: A critical review and synthesis of research on cyberbullying victimization. *Computers in Human Behavior* 26, no. 3, 277–287. doi:10:1016/j.chb.2009:11.014.

4. Lenhart, L., Madden, M., Smith, A., Purcell, K., Zickuhr, K., and Rainie, L. (2011). *Teens, kindness and cruelty on social network sites*. Retrieved from http://pewinternet.org/Reports/2011/Teens-and-social-media.aspx, pp. 26–27.

5. de Balzac, H. (2010). *Father Goriot*. (E. Marriage, Trans). Project Gutenberg ebook. Retrieved from http://www.gutenberg.org/files/1237/1237-h/1237-h.htm. (Original work published 1835.)

6. Volk, A. A., Camilleri, J. A., Dane, A. V., and Marini, Z. A. (2012). Is adolescent bullying an evolutionary adaptation? *Aggressive Behavior* 38, no. 3, 222–238. doi:10:1002/ab.21418.

7. Olweus, D. (2012). Cyberbullying: An overrated phenomenon? *European Journal of Developmental Psychology* 9, no. 5, 520–538. doi:10:1080/17405629:2012:682358, p. 529.

8. Bonanno, R. A., and Hymel, S. (2013). Cyber bullying and internalizing difficulties: Above and beyond the impact of traditional forms of bullying. *Journal of Youth and Adolescence* 42, no. 5, 685–697. doi:10:1007/s10964-013-9937-1.

9. Pornari, C. D., and Wood, J. (2010). Peer and cyber aggression in secondary school students: The role of moral disengagement, hostile attribution bias, and outcome expectancies. *Aggressive Behavior* 36, no. 2, 81–94. doi:10:1002/ab.20336.

10. Bandura, A. (1986). Social foundation of thought and action: A social cognitive theory. Englewood Cliffs, NJ: Prentice Hall.

11. Pornari and Wood, 2010.

12. Perren, S., and Gutzwiller-Helfenfinger, E. (2012). Cyberbullying and traditional bullying in adolescence: Differential roles of moral disengagement, moral emotions, and moral values. *European Journal of Developmental Psychology* 9, no. 2, 195–209. doi:10:1080/17405629:2011:643168.

13. Robson, C., and Witenberg, R. T. (2013). The influence of moral disengagement, morally based self-esteem, age, and gender on traditional and cyber

bullying. *Journal of School Violence* 12, 211–231. doi:10:1080/15388220:2012: 762921.

14. Hardaker, C. (2010). Trolling in asynchronous computer-mediated communication: From user discussions to academic definitions. *Journal of Politeness Research: Language, Behavior, Culture* 6, no. 2, 215–242. doi:10:1515/jplr.2010:011.

15. Carey, T. (September 24, 2011). "Help me, mummy. It's hot here in hell": A special investigation into the distress of grieving families caused by the sick Internet craze "trolling." Retrieved from http://www.dailymail.co.uk/news/article-2041193/Internet-trolling-Investigation-distress-grieving-families-caused-trolls.html.

16. Quoted in Paton, G. (October 2010). Facebook "encourages children to spread gossip and insults." Retrieved from http://www.telegraph.co.uk/education/educationnews/8067093/Facebook-encourages-children-to-spread-gossip-and-insults.html.

17. Jackson, L. A., and Wang, J. L. (2013). Cultural differences in social networking site use: A comparative study of China and the United States. *Computers in Human Behavior* 29, no. 3, 910–921. doi:10:1016/j.chb.2012:11.024.

18. Anderson, B., Fagan, P., Woodnutt, T., and Chamorro-Prezumic, T. (2012). Facebook psychology: Popular questions answered by research. *Psychology of Popular Media Culture* 1, no. 1, 23–37. doi:10:1037/a0026452.

19. Huang, C. (June 6, 2011). Facebook and Twitter key to Arab Spring uprisings: Report. Retrieved from http://www.thenational.ae/news/uae-news/facebook-and-twitter-key-to-arab-spring-uprisings-report.

20. Waldorf, L. (2012). White noise: Hearing the disaster. *Journal of Human Rights Practice* 4, no. 3, 469–474. doi:10:1093/jhuman/hus025.

21. Flores, A., and James, C. (2013). Morality and ethics behind the screen: Young people's perspectives on digital life. *New Media & Society* 15, 834–852. doi:10:1177/1461444812462842.

22. Donne, J. (1839). Devotions upon emergent occasions. In H. Alfred (Ed.), The works of John Donne, pp. 574–575. Retrieved from http://www.luminarium.org/sevenlit/donne/meditation17.php.

CHAPTER 13. THE *SOMETHING* ABOUT VIDEOGAMES

1. Entertainment Software Association. (2013). *Essential facts about the computer and videogame industry*. Retrieved from http://www.theesa.com/facts/pdfs/esa_ef_2013.pdf.

2. Entertainment Software Association, 2013.

3. Lazzaro, N. (2004). *Why we play games: Four keys to more emotion without story*. Retrieved from http://www.xeodesign.com/xeodesign_whyweplaygames.pdf.

4. D'Angelo, W. (April 23, 2012). Top 10 in sales—first person shooters [blog post]. Retrieved from http://www.vgchartz.com/article/250080/top-10-in-sales-first-person-shooters.

5. Demetrovics, Z., Urbán, R., Nagygyörgy, K., Farkas, J., Zilahy, D., Mervó, B., . . . and Harmath, E. (2011). Why do you play? The development of the motives for online gaming questionnaire (MOGQ). *Behavior Research Methods* 43, no. 3, 814–825. doi:10:3758/s13428–011–0091-y. Yee, N. (2006). Motivations for play in online games. *CyberPsychology & Behavior* 9, no. 6, 772–775. doi:10:1089/cpb.2006:9.772.

6. Kuss, D. J., and Griffiths, M. D. (2012). Internet gaming addiction: A systematic review of empirical research. *International Journal of Mental Health and Addiction* 10, no. 2, 278–296. doi:10:1007/s11469-011-9138-5.
7. Hopson, J. (April 27, 2001). Behavioral game design. Retrieved from http://www.gamasutra.com/view/feature/3085/behavioral_game_design.php.
8. Yee, N. (2002). Facets: 5 motivation factors for why people play MMORPGs. Retrieved from http://www.nickyee.com/facets/home.html.
9. The reason that Internet addiction studies are included is because cyberpsychologists are currently unclear as to whether videogame addiction is the same thing as Internet addiction. Ten years ago, when lots of these studies were being done, the two were conflated, as researchers assumed Internet addiction was the same thing as videogame addiction, or that videogame addiction was a manifestation of Internet addiction. Thus, researchers looked at Internet addiction in samples of excessive gamers. We would be unable to gain a complete picture of the current literature without also including Internet addiction.
10. Kuss and Griffiths, 2012. Hur, M. H. (2012). Current trends of Internet addiction disorder research: A review of 2000–2008 Korean academic journal articles. *Asia Pacific Journal of Social Work and Development* 22, no. 3, 187–201. doi:10:1080/02185385:2012:691718.
11. Back in 1998, and therefore predating widespread gaming on the Internet, Dr. Kimberly Young modified preexisting criteria used to diagnose pathological gambling in order to suggest that pathological Internet use shared similar features: preoccupation, tolerance, withdrawal, failure to control, longer use than intended, functional impairment, lying, and escape (Young, K. S. [1998]. Internet addiction: The emergence of a new clinical disorder. *CyberPsychology & Behavior* 1, no. 3, 237–244. doi:10:1089/cpb.1998: 1.237). As the director of the Center for Internet Addiction Recovery, Young takes the view that "Internet addicts suffer from emotional problems such as depression and anxiety-related disorders and often use the fantasy world of the Internet to psychologically escape unpleasant feelings or stressful situations" (Young, K. [March 15, 2012]. FAQs. Retrieved from http://netaddiction .com/faqs). Consequences include regular loss of sleep, changes to diet, relationship difficulties, damage to real-world social life, loss of income or employment, poorer academic performance, irritability or anxiety when not using the Internet, and an inability to cut back or stop Internet use. As if that weren't enough, excessive use has been linked to higher levels of hostility, stress, loneliness, depression, and increased suicidal thoughts (Ko, C. H., Yen, J. Y., Yen, C. F., Chen, C. S., and Chen, C. C. [2012]. The association between Internet addiction and psychiatric disorder: A review of the literature. *European Psychiatry* 27, no. 1, 1–8. doi:10:1016/j.eurpsy.2010:04.011). Another lobbyist is David Greenfield (no relation), who heads up the Center for Internet Behavior in Connecticut and is the author of *Virtual Addiction* (Greenfield, D. N. [1999]. *Virtual Addiction: Help for Netheads, Cyberfreaks, and Those Who Love Them*. Oakland, CA: New Harbinger). Greenfield believes that some services available over the Internet offer an unprecedented and alluring cocktail of stimulating content, ease of access, convenience, low cost, visual stimulation, autonomy, and anonymity, all of which contribute to a highly psychoactive experience. Defining "psychoactive" as something that alters mood and potentially impacts behavior, Greenfield claims that

online sex, gaming, gambling, and shopping can all produce a mood-altering effect, suggesting that a wide variety of Internet activities could all be lumped together as "addictive."

12. Signs of gaming addiction in adults. (n.d.). Retrieved from http://www.video-game-addiction.org/internet-addictions-adults.html.

13. Starcevic, V. (2013). Is Internet addiction a useful concept? *Australian and New Zealand Journal of Psychiatry* 47, no. 1, 16–19. doi:10:1177/0004867412461693.

14. In the United States, a review of excessive Internet use in youths showed prevalence ranging between 0 and 26.3 percent (Moreno, M. A., Jelenchick, L., Cox, E., Young, H., and Christakis, D. A. [2011]. Problematic Internet use among U.S. youth: A systematic review. *Archives of Pediatrics & Adolescent Medicine* 165, no. 9, 797–805. doi:10:1001/archpediatrics.2011:58). Meanwhile, in Hong Kong, Daniel Shek, a psychology professor at the Polytechnic University and an expert on "healthy parenting," came up with an estimated prevalence of Internet addiction of 20 percent in Chinese adolescents (Shek, D. T., Tang, V. M., and Lo, C. Y. [2008]. Internet addiction in Chinese adolescents in Hong Kong: Assessment, profiles, and psychosocial correlates. *Scientific World Journal* 8, 776–787. doi:10:1100/tsw.2008:104). In contrast, in a European study Konstantinos Siomos, a child and adolescent psychiatrist and chair of the Hellenic Society for the Study of Internet Addiction Disorder, surveyed more than two thousand Greek teenagers, again sampled using a diagnostic questionnaire of Internet addiction. Siomos found that the prevalence of Internet addiction was 8.2 percent (Siomos, K. E., Dafouli, E. D., Braimiotis, D. A., Mouzas, O. D., and Angelopoulos, N. V. [2008]. Internet addiction among Greek adolescent students. *CyberPsychology & Behavior* 11, no. 6, 653–657. doi:10:1089/cpb.2008:0088). A review of the publications produced until the end of 2009 showed, from different studies mainly conducted in the Far East, highly varied prevalence rates of Internet addiction (Ko et al., 2012).

15. Gentile, D. (2009). Pathological video-game use among youth ages 8 to 18: A national study. *Psychological Science* 20, no. 5, 594–602. doi:10:1111/j.1467-9280:2009:02340.x.

16. Kuss and Griffiths, 2012.

17. Gentile, D. A., Choo, H., Liau, A., Sim, T., Li, D., Fung, D., and Khoo, A. (2011). Pathological videogame use among youths: A two-year longitudinal study. *Pediatrics* 127, no. 2, 319–329. doi:10:1542/peds.2010–1353.

18. Weinstein, A., and Lejoyeux, M. (2010). Internet addiction or excessive Internet use. *American Journal of Drug and Alcohol Abuse* 36, no. 5, 277–283. doi:10:3109/00952990:2010:491880.

19. Weinstein, A. M. (2010). Computer and videogame addiction: A comparison between game users and non-game users. *American Journal of Drug and Alcohol Abuse* 36, no. 5, 268–276. doi:10:3109/00952990:2010:491879.

20. The ventral striatum is the lower part and the dorsal striatum the upper part of a large area, the striatum, which occupies a central part of the brain in all mammals. In higher animals the striatum is subdivided into the upper part, called the putamen, and the lower, the caudate nucleus. Within the ventral striatum/caudate nucleus, another area, the nucleus accumbens, is a dopamine-rich region closely linked to drug addiction. The striatum is also directly and reciprocally connected to the substantia nigra, the primary site of degeneration in Parkinson's disease, a movement disorder. There is no

single "function" for the striatum, which has diffuse and complex anatomical connections and which has been related to a variety of processes ranging
from reward to movement.

21. Kühn, S., Romanowski, A., Schilling, C., Lorenz, R., Mörsen, C., Seiferth, N., . . . and Gallinat, J. (2011). The neural basis of gaming. *Translational Psychiatry* 1, no. 11, e53. doi:10:1038/tp.2011:53.
22. Linnet, J., Peterson, E., Doudet, D. J., Gjedde, A., and Moller, A. (2010). Dopamine release in ventral striatum of pathological gamblers losing money. *Acta Psychiatrica Scandinavica* 112, no. 4, 326–333. doi:10:1111/j.1600-0447:2010:01591.x.
23. Kühn et al., 2011.
24. Erickson, K. I., Boot, W. R., Basak, C., Neider, M. B., Prakash, R. S., Voss, M. W., . . . and Kramer, A. F. (2010). Striatal volume predicts level of video-game skill acquisition. *Cerebral Cortex* 20, no. 11, 2522–2530. doi:10:1093/cercor/bhp293.
25. Drevets, W. C., Price, J. C., Kupfer, D. J., Kinahan, P. E., Lopresti, B., Holt, D., and Mathis, C. (1999). PET measures of amphetamine-induced dopamine release in ventral versus dorsal striatum. *Neuropsychopharmacology* 21, no. 6, 694–709. doi:10:1016/S0893-133X(99)00079-2.
26. Robbins, T. W., and Everitt, B. J. (1992). Functions of dopamine in the dorsal and ventral striatum. *Seminars in Neuroscience* 4, no. 2, 119–127. Retrieved from http://www.sciencedirect.com/science/article/pii/104457659290010Y. MacDonald, P. A., MacDonald, A. A., Seergobin, K. N., Tamjeedi, R., Ganjavi, H., Provost, J. S., and Monchi, O. (2011). The effect of dopamine therapy on ventral and dorsal striatum-mediated cognition in Parkinson's disease: Support from functional MRI. *Brain* 134, no. 5, 1447–1463. doi:10:1093/brain/awr075.
27. Koepp, M. J., Gunn, R. N., Lawrence, A. D., Cunningham, V. J., Dagher, A., Jones, T., . . . and Grasby, P. M. (1998). Evidence for striatal dopamine release during a videogame. *Nature* 393, no. 6682, 266–267. doi:10:1038/30498.
28. Metcalf, O., and Pammer, K. (2014). Sub-types of gaming addiction: Physiological arousal deficits in addicted gamers differ based on preferred genre. *European Addiction Research* 20, no. 1, 23–32. doi:10:1159/000349907.
29. Han, D. H., Lee, Y. S., Yang, K. C., Kim, E. Y., Lyoo, I. K., and Renshaw, P. F. (2007). Dopamine genes and reward dependence in adolescents with excessive Internet videogame play. *Journal of Addiction Medicine* 1, no. 3, 133–138. doi:10:1097/ADM.0b013e31811f465f.
30. Lush, T. (August 29, 2011). At war with *World of Warcraft:* An addict tells his story. Retrieved from http://www.theguardian.com/technology/2011/aug/29/world-of-warcraft-video-game-addict.
31. Personal communication, March 8, 2013.
32. King, D. L., Delfabbro, P. H., and Griffiths, M. D. (2011). The role of structural characteristics in problematic videogame play: An empirical study. *International Journal of Mental Health and Addiction* 9, no. 3, 320–333. doi:10:1007/s11469-010-9289-y.
33. Lazzaro, 2004.
34. Trepte, S. and Reinecke, L. (2010). Avatar creation and videogame enjoyment: Effects of life-satisfaction, game competitiveness, and identification with the avatar. *Journal of Media Psychology: Theories, Methods, and Applications* 22, no. 4, 171–184. doi:10:1027/1864–1105/a000022.
35. Bavelier, D., Green, C. S., and Dye, M. W. (2010). Children, wired: For better

and for worse. *Neuron* 67, no. 5, 692–701. doi:10:1016/j.neuron.2010:08.035, p. 698.

36. Nunneley, S. (April 30, 2013). Guardian analysis of top 50 games sold in 2012 found "more than half contain violent content labels" [blog post]. Retrieved from http://www.vg247.com/2013/04/30/guardian-analysis-of-top-50-games-sold-in-2012-found-more-than-half-contain-violent-content-labels.

CHAPTER 14. VIDEOGAMES AND ATTENTION

1. Dowd, M. (December 6, 2011). Silence is golden. Retrieved from http://www.nytimes.com/2011/12/07/opinion/dowd-silence-is-golden.html?_r=0.

2. Christakis, D. A., Zimmerman, F. J., DiGiuseppe, D. L., and McCarty, C. A. (2004). Early television exposure and subsequent attentional problems in children. *Pediatrics* 113, 708–713. Retrieved from http://pediatrics.aappublications.org/content/113/4/708.short.

3. Some experts think that gaming is indeed more detrimental than TV. In his recent review, psychologist Paul Howard-Jones points to the difference between the two media in terms of the degree of personal involvement and interactivity. He concludes: "In terms of content . . . it seems the Internet leisure activities popular with children, such as games, might not teach the types of attentional capabilities required for 'paying attention' in the classroom and other contexts. Given the additional interactivity and the levels of psychological and cognitive engagement they can provide, a case can be made that some Internet activities (such as games) might pose a greater threat to some attentional abilities other than television" (Howard-Jones, P. [2011]. *The impact of digital technologies on human wellbeing: Evidence from the sciences of mind and brain*. Retrieved from http://www.nominettrust.org.uk/sites/default/files/NT%20SoA%20-%20The%20impact%20of%20digital%20technologies%20on%20human%20wellbeing.pdf).

4. Swing, E. L., Gentile, D. A., Anderson, C. A., and Walsh, D. A. (2010). Television and videogame exposure and the development of attention problems. *Pediatrics* 126, no. 2, 214–221. doi:10:1542/peds.2009–1508.

5. Swing et al., 2010.

6. Gentile, D. A., Swing, E. L., Lim, C. G., and Khoo, A. (2012). Videogame playing, attention problems, and impulsiveness: Evidence of bidirectional causality. *Psychology of Popular Media Culture* 1, no. 1, 62. doi:10:1037/a0026969.

7. McKinley, R. A., McIntire, L. K., and Funke, M. A. (2011). Operator selection for unmanned aerial systems: Comparing videogame players and pilots. *Aviation, Space and Environmental Medicine* 82, no. 6, 635–642. doi:10:3357/ASEM.2958:2011.

8. Appelbaum, L. G., Cain, M. S., Darling, E. F., and Mitroff, S. R. (2013). Action videogame playing is associated with improved visual sensitivity, but not alterations in visual sensory memory. *Attention, Perception, & Psychophysics* 75, no. 6, 1161–1167. doi: 10:3758/s13414-013-0472-7.

9. Quoted in Gamers really do see more. (June 11, 2013). *Duke Today*. Retrieved from http://today.duke.edu/2013/06/vidvision.

10. Boot, W. R., Blakely, D. P., and Simons, D. J. (2011). Do action videogames improve perception and cognition? *Frontiers in Psychology* 2, 1–6. doi:10:3389/fpsyg.2011:00226.

11. Bavelier, D., Green, C. S., Pouget, A., and Schrater, P. (2012). Brain plasticity through the life span: Learning to learn and action videogames. *Annual Review of Neuroscience* 35, 391–416. doi:10:1146/annurev-neuro-060909 -152832. Castel, A. D., Pratt, J., and Drummond, E. (2005). The effects of action videogame experience on the time course of inhibition of return and the efficiency of visual search. *Acta Psychologica* 119, no. 2, 217–230. doi:10:1016/j.actpsy.2005:02.004. Donohue, S. E., Woldorff, M. G., and Mitroff, S. R. (2010). Videogame players show more precise multisensory temporal processing abilities. *Attention, Perception & Psychophysics* 72, no. 4, 1120–1129. doi:10:3758/APP.72:4.1120. Dye, M. W. G., Green, C. S., and Bavelier, D. (2009). Increasing speed of processing with action videogames. *Current Directions in Psychological Science* 18, no. 6, 321–326. doi:10:1111/ j.1467–8721:2009:01660.x. Feng, J., Spence, I., and Pratt, J. (2007). Playing an action videogame reduces gender differences in spatial cognition. *Psychological Science* 18, no. 10, 850–855. doi:10:1111/j.1467–9280:2007:01990.x. Green, C. S., and Bavelier, D. (2003). Action videogame modifies visual selective attention. *Nature* 423, no. 6939, 534–537. Retrieved from http://www .nature.com/nature/journal/v423/n6939/abs/nature01647.html. Green, C. S., and Bavelier, D. (2007). Action-video-game experience alters the spatial resolution of vision. *Psychological Science* 18, no. 1, 88–94. doi:10:1111/ j.1467–9280:2007:01853.x. Green, C. S., and Bavelier, D. (2012). Learning, attentional control, and action videogames. *Current Biology* 22, no. 6, R197–R206. doi:10:1016/j.cub.2012:02.012. Green, C. S., Pouget, A., and Bavelier, D. (2010). Improved probabilistic inference as a general learning mechanism with action videogames. *Current Biology* 20, 1573–1579. doi:10:1016/j.cub.2010:07.040. Hubert-Wallander, B., Green, C. S., and Bavelier, D. (2011). Stretching the limits of visual attention: The case of action videogames. *Wiley Interdisciplinary Reviews: Cognitive Science* 2, no. 2, 222– 230. doi:10:1002/wcs.116. Subrahmanyam, K., and Greenfield, P. M. (1994). Effect of videogame practice on spatial skills in girls and boys. *Journal of Applied Developmental Psychology* 15, no. 1, 13–32. doi:10:1016/0193-3973(94) 90004-3.

12. Green and Bavelier (2003), p. 536. The subjects were between eighteen and twenty-three years old and were divided into two separate groups. The gamers were already familiar with playing action videogames such as Grand Theft Auto for a minimum of one hour a day, at least four days a week, during the previous six months. The second group, the nonplayers, had had little or no videogame usage in the previous six months.

13. See citations in n. 11.

14. Rosser, J. C. J., Lynch, P. J., Cuddihy, L., Gentile, D. A., Klonsky, J., Merrell, R., and Curet, M. (2007). The impact of videogames on training surgeons in the 21st century. *Archives of Surgery* 142, no. 2, 181–186. Retrieved from http://cat .inist.fr/?aModele=afficheN&cpsidt=18510967.

15. What is Big Brain Academy? (n.d.). Retrieved from http://www.bigbrain academy.com/ds/what/index.html.

16. Boot, Blakely, and Simons, 2011.

17. Han, D. H., Renshaw, P. F., Sim, M. E., Kim, J. I., Arenella, L. S., and Lyoo, I. K. (2008). The effect of Internet videogame play on clinical and extrapyramidal symptoms in patients with schizophrenia. *Schizophrenia Research* 103, nos. 1–3, 338–340. doi:10:1016/j.schres.2008:01.026.

18. Bavelier, D., Green, C. S., Han, D. H., Renshaw, P. F., Merzenich, M. M., and Gentile, D. A. (2011). Brains on videogames. *Nature Reviews Neuroscience* 12, no. 12, 763–768. doi:10:1038/nrn3135.
19. Walshe, D. G., Lewis, E. J., Kim, S. I., O'Sullivan, K., and Wiederhold, B. K. (2003). Exploring the use of computer games and virtual reality in exposure therapy for fear of driving following a motor vehicle accident. *CyberPsychology & Behavior* 6, no. 3, 329–334. doi:10:1089/109493103322011641.
20. Fernández-Aranda, F., Jiménez-Murcia, S., Santamaría, J. J., Gunnard, K., Soto, A., Kalapanidas, E., . . . and Penelo, E. (2012). Videogames as a complementary therapy tool in mental disorders: PlayMancer, a European multicentre study. *Journal of Mental Health* 21, no. 4, 364–374. doi:10:3109/096382 37:2012:664302.
21. Gambotto-Burke, A. (August 13, 2011). Hi-tech stimuli help to dull the pain. Retrieved from http://www.theaustralian.com.au/news/health-science/hi -tech-stimuli-help-to-dull-the-pain/story-e6frg8y6–1226113730661.
22. Coyne, S. M., Padilla-Walker, L. M., Stockdale, L., and Day, R. D. (2011). Game on . . . girls: Associations between co-playing videogames and adolescent behavioral and family outcomes. *Journal of Adolescent Health* 49, no. 2, 160–165. doi:10:1016/j.jadohealth.2010:11.249.
23. Bessière, K., Seay, A. F., and Kiesler, S. (2007). The ideal elf: Identity exploration in *World of Warcraft*. *CyberPsychology & Behavior* 10, no. 4, 530–535. doi:10:1089/cpb.2007:9994.
24. Xanthopoulou, D., and Papagiannidis, S. (2012). Play online, work better? Examining the spillover of active learning and transformational leadership. *Technological Forecasting and Social Change* 79, no. 7, 1328–1339. doi:10:1016/ j.techfore.2012:03.006.
25. Bailey, K., West, R., and Anderson, C. A. (2010). A negative association between videogame experience and proactive cognitive control. *Psychophysiology* 47, no. 1, 34–42. doi:10:1111/j.1469–8986:2009:00925.x.
26. Braver, T. S., Gray, J. R., and Burgess, G. C. (2007). Explaining the many varieties of working memory variation: Dual mechanisms of cognitive control. In A. Conway, C. Jarrold, M. J. Kane, A. Miyake, and J. N. Towse (Eds.), *Variation in working memory,* pp. 76–106. Oxford: Oxford University Press.
27. Bailey, West, and Anderson, 2010.
28. Ventura, M., Shute, V., and Zhao, W. (2013). The relationship between videogame use and a performance-based measure of persistence. *Computers & Education* 60, no. 1, 52–58. doi:10:1016/j.compedu.2012:07.003.
29. ADHD and ADD (attention deficit disorder) are general terms frequently used to describe individuals who have attention deficit hyperactivity disorder or, as the abbreviated name suggests, attention problems without the hyperactive and impulsive behaviors. The terms ADHD and ADD are often used interchangeably both for those who do have symptoms of hyperactivity and impulsiveness and for those who do not. For our purposes we can consider the two together under the same umbrella terminology relating to attentional problems.
30. Howard-Jones, P. (2011). *The impact of digital technologies on human wellbeing: Evidence from the sciences of mind and brain.* Retrieved from http://www .nominettrust.org.uk/sites/default/files/NT%20SoA%20-%20The%20impact %20of%20digital%20technologies%20on%20human%20wellbeing.pdf, p. 52.

31. Parkes, A., Sweeting, H., Wight, D., Henderson, M. (2013). Do television and electronic games predict children's psychosocial adjustment? Longitudinal research using the UK Millennium Cohort Study. *Archives of Disease in Childhood* 98, no. 5, 341–348. doi:10:1136/archdischild-2011-301508.

32. Researchers use terms such as "excessive," "obsessive," "compulsive," "pathological," "harmful," and "addictive" when referring to Internet use that results in harmful consequences to a user and an inability to control Internet use, as "Internet addiction" is not a formally recognized disorder. These terms are used throughout to refer to the same behavior—an inability to cut back or control Internet use, despite significant negative consequences.

33. Collins, E., Freeman, J., and Chamarro-Premuzic, T. (2012). Personality traits associated with problematic and non-problematic massively multiplayer online role playing game use. *Personality and Individual Differences* 52, no. 2, 133–138. doi:10:1016/j.paid.2011:09.015.

34. Ha, J. H., Yoo, H. J., Cho, I. H., Chin, B., Shin, D., and Kim, J. H. (2006). Psychiatric comorbidity assessed in Korean children and adolescents who screen positive for Internet addiction. *Journal of Clinical Psychiatry* 67, no. 5, 821–826. doi:10:4088/JCP.v67n0517.

35. Yen, J. Y., Ko, C. H., Yen, C. F., Wu, H. Y., and Yang, M. J. (2007). The comorbid psychiatric symptoms of Internet addiction: Attention deficit and hyperactivity disorder (ADHD), depression, social phobia, and hostility. *Journal of Adolescent Health* 41, 93–98. doi:10:1016/j.jadohealth.2007:02.002.

36. Swing et al., 2010.

37. Chan, P. A., and Rabinowitz, T. (2006). A cross-sectional analysis of videogames and attention deficit hyperactivity disorder symptoms in adolescents. *Annals of General Psychiatry* 5, no. 16, 5–16. doi:10:1186/1744–859X-5-16, p. 4.

38. Bioulac, S., Arfi, L., and Bouvard, M. P. (2008). Attention deficit/hyperactivity disorder and videogames: A comparative study of hyperactive and control children. *European Psychiatry* 23, no. 2, 134–141. doi:10:1016/j.eurpsy.2007:11.002.

39. Doward, J., and Craig, E. (May 6, 2012). Ritalin use for ADHD children soars fourfold. Retrieved from http://www.theguardian.com/society/2012/may/06/ritalin-adhd-shocks-child-psychologists.

40. Zuvekas, S. H., and Vitiello, B. (2012). Stimulant medication use in children: A 12-year perspective. *American Journal of Psychiatry* 169, no. 2, 160–166. doi:10:1176/appi.ajp.2011:11030387.

41. Hollingworth, S. A., Nissen, L. M., Stathis, S. S., Siskind, D. J., Varghese, J. M., and Scott, J. G. (2011). Australian national trends in stimulant dispensing: 2002–2009. *Australian and New Zealand Journal of Psychiatry* 45, no. 4, 332–336. doi:10:3109/00048674:2010:543413.

42. A major thought to flag here, however, is that whatever findings may emerge from one group of subjects, we must always be aware of possible cultural factors making an all-important difference. For example, a Korean scientist, Seok Young Moon, all too aware of the issue, concluded: "While investigating whether teachers' and parents' perspectives on ADHD are influenced by culture, I found that cultural influence plays an important role: In Korea, according to Confucianism, parents and teachers tend to focus more on children's academic achievement and take children's distractive behaviors as a negative reflection on themselves and their authority. Korean teachers and parents try to take personal responsibility for children's distractive be-

haviors, and have negative attitudes toward medication because the medication does not help to increase academic improvement. U.S. parents and teachers, influenced by western culture's focus on independence, tend not to take personal responsibility for the children's behaviors but to focus more on children's current problems and treatment. U.S. parents and teachers did not mind a third party's engagement in dealing with children with ADHD and their behaviors. U.S. parents were more positive about medical treatments because medication helps to reduce children's distractive behaviors" (Moon, S. Y. [n.d.]. Cultural perspectives on attention deficit hyperactivity disorder: A comparison between Korea and the United States. *Journal of International Business and Cultural Studies,* 1–11. Retrieved from http://ww.aabri .com/manuscripts/11898.pdf).

43. Turner, D. C., Robbins, T. W., Clark, L., Aron, A. R., Dowson, J., and Sahakian, B. J. (2003). Cognitive enhancing effects of modafinil in healthy volunteers. *Psychopharmacology* 165, no. 3, 260–269. doi:10:1007/s00213-002-1250-8.

44. Han, D. H., Lee, Y. S., Na, C., Ahn, J. Y., Chung, U. S., Daniels, M. A., . . . and Renshaw, P. F. (2009). The effect of methylphenidate on Internet videogame play in children with attention-deficit/hyperactivity disorder. *Comprehensive Psychiatry* 50, no. 3, 251–256. doi:10:1016/j.comppsych.2008:08.011, p. 251.

45. Howard-Jones, 2011.

CHAPTER 15. VIDEOGAMES, AGGRESSION, AND RECKLESSNESS

1. Fifty-six children ages ten to thirteen named other children who displayed certain forms of physical and verbal aggressive behavior that day, as well as evaluating the intentions of these aggressive behaviors, if any. Polman, H., De Castro, B. O., and van Aken, M. A. (2008). Experimental study of the differential effects of playing versus watching violent videogames on children's aggressive behavior. *Aggressive Behavior* 34, no. 3, 256–264. doi:10:1002/ ab.20245.

2. Griffiths, M. (1999). Violent videogames and aggression: A review of the literature. *Aggression and Violent Behavior* 4, no. 2, 203–212. doi:10:1016/S1359– 1789(97)00055-4. Dill, K. E., and Dill, J. C. (1999). Videogame violence: A review of the empirical literature. *Aggression and Violent Behavior* 3, no. 4, 407–428. doi:10:1016/S1359–1789(97)00001-3. Anderson, C. A., and Dill, K. E. (2000). Videogames and aggressive thoughts, feelings, and behavior in the laboratory and in life. *Journal of Personality and Social Psychology* 78, no. 4, 772–790. doi:10:1037/0022–3514:78.4:772.

3. Konijn, E. A., Nije Bijvank, M., and Bushman, B. J. (2007). I wish I were a warrior: The role of wishful identification in the effects of violent videogames on aggression in adolescent boys. *Developmental Psychology* 43, no. 4, 1038–1044. doi:10:1037/0012–1649:43.4:1038.

4. Weaver, A. J., and Lewis, N. (2012). Mirrored morality: An exploration of moral choice in videogames. *Cyberpsychology, Behavior, and Social Networking* 15, no. 11, 610–614. doi:10:1089/cyber.2012:0235.

5. Weaver and Lewis, 2012.

6. Anderson, C. A. (2003). Violent videogames: Myths, facts, and unanswered

questions. Retrieved from http://www.apa.org/science/about/psa/2003/10/anderson.aspx.

7. Bavelier, D., Green, C. S., Han, D. H., Renshaw, P. F., Merzenich, M. M., and Gentile, D. A. (2011). Brains on videogames. *Nature Reviews Neuroscience* 12, no. 12, 763–768. doi:10:1038/nrn3135, p.765.

8. These effects were also more marked for men than for women. Given the far higher levels of testosterone in the male body, a tendency for aggression will be much more easily realized (Hasan, Y., Bègue, L., and Bushman, B. J. [2012]. Viewing the world through "blood-red tinted glasses": The hostile expectation bias mediates the link between violent videogame exposure and aggression. *Journal of Experimental Social Psychology* 48, no. 4, 953–956. doi:10: 1016/j.jesp.2011:12.019).

9. Ferguson, C. J. (2009). Violent videogames: Dogma, fear, and pseudoscience. *Skeptical Inquirer* 33, 38–43. Retrieved from http://www.tamiu.edu/~cferguson/skeptinq.pdf.

10. Olson, C. K. (2010). Children's motivations for videogame play in the context of normal development. *Review of General Psychology* 14, no. 2, 180–187. doi:10:1037/a0018984.

11. Anderson, C. A., Sakamoto, A., Gentile, D. A., Ihori, N., Shibuya, A., Yukawa, S., . . . and Kobayashi, K. (2008). Longitudinal effects of violent videogames on aggression in Japan and the United States. *Pediatrics* 122, no. 5, e1067–e1072. doi:10:1542/peds.2008-1425.

12. Möller, I., and Krahé, B. (2009). Exposure to violent videogames and aggression in German adolescents: A longitudinal analysis. *Aggressive Behavior* 35, no. 1, 75–89. doi:10:1002/ab.20290.

13. Wallenius, M., and Punamäki, R. L. (2008). Digital game violence and direct aggression in adolescence: A longitudinal study of the roles of sex, age, and parent-child communication. *Journal of Applied Developmental Psychology* 29, no. 4, 286–294. doi:10:1016/j.appdev.2008:04.010.

14. Anderson, C. A., Shibuya, A., Ihori, N., Swing, E. L., Bushman, B. J., Sakamoto, A., . . . and Saleem, M. (2010). Violent videogame effects on aggression, empathy, and prosocial behavior in eastern and western countries: A meta-analytic review. *Psychological Bulletin* 136, no. 2, 151. doi:10:1037/a0018251.

15. Ferguson, C. J., and Kilburn, J. (2010). Much ado about nothing: The misestimation and overinterpretation of violent videogame effects in Eastern and Western nations: Comment on Anderson et al. (2010). *Psychological Bulletin* 136, no. 2, 174–178. doi:10:1037/a0018566.

16. Bushman, B. J., Rothstein, H. R., and Anderson, C. A. (2010). Much ado about something: Violent videogame effects and a school of red herring: Reply to Ferguson and Kilburn (2010). *Psychological Bulletin,* 136, no. 2, 182–187. doi:10:1037/a0018718.

17. Ferguson, 2009. Gunter, W. D., and Daly, K. (2012). Causal or spurious: Using propensity score matching to detangle the relationship between violent videogames and violent behavior. *Computers in Human Behavior* 28, no. 4, 1348–1355. doi:10:1016/j.chb.2012:02.020.

18. Carnagey, N. L., Anderson, C. A., and Bushman, B. J. (2007). The effect of videogame violence on physiological desensitization to real-life violence. *Journal of Experimental Social Psychology* 43, no. 3, 489–496. doi:10:1016/j.jesp.2006:05.003.

19. Carnagey, Anderson, and Bushman, 2007.
20. Bushman, B. J., and Anderson, C. A. (2009). Comfortably numb desensitizing effects of violent media on helping others. *Psychological Science* 20, no. 3, 273–277. doi:10:1111/j.1467–9280:2009:02287.x.
21. Mathiak, K., and Weber, R. (2006). Toward brain correlates of natural behavior: fMRI during violent videogames. *Human Brain Mapping* 27, no. 12, 948–956. doi:10:1002/hbm.20234.
22. Weber, R., Ritterfeld, U., and Mathiak, K. (2006). Does playing violent videogames induce aggression? Empirical evidence of a functional magnetic resonance imaging study. *Media Psychology* 8, no. 1, 39–60. doi:10:1207/S1532785XMEP0801_4.
23. Llinás, R. R., and Paré, D. (1991). Of dreaming and wakefulness. *Neuroscience* 44, no. 3, 521–535. doi.org/10:1016/0306–4522(91)90075-Y.
24. Nunneley, S. (April 30, 2013). Guardian analysis of top 50 games sold in 2012 found "more than half contain violent content labels" [blog post]. Retrieved from http://www.vg247.com/2013/04/30/guardian-analysis-of-top-50-games-sold-in-2012-found-more-than-half-contain-violent-content-labels.
25. Davis, C., Levitan, R. D., Muglia, P., Bewell, C., and Kennedy, J. L. (2004). Decision-making deficits and overeating: A risk model for obesity. *Obesity Research* 12, no. 6, 929–935. doi:10:1038/oby.2004:113. Pignatti, R., Bertella, L., Albani, G., Mauro, A., Molinari, E., and Semenza, C. (2006). Decision-making in obesity: A study using the Gambling Task. *Eating and Weight Disorders: EWD* 11, no. 3, 126–132. Retrieved from http://www.ncbi.nlm.nih.gov/pubmed/17075239.
26. Oltmanns, T. F. (1978). Selective attention in schizophrenic and manic psychoses: The effect of distraction on information processing. *Journal of Abnormal Psychology* 87, no. 2, 212–225. Retrieved from http://www.ncbi.nlm.nih.gov/pubmed/649860. Parsons, B., Gandhi, S., Aurbach, E. L., Williams, N., Williams, M., Wassef, A., and Eagleman, D. M. (2012). Lengthened temporal integration in schizophrenia. *Neuropsychologia* 51, no. 2, 372–376. doi:10:1016/j.neuropsychologia.2012:11.008.
27. Kasanin, J. S. (Ed.) (1944). *Language and thought in schizophrenia*. Berkeley, CA: University of California Press. In the case of schizophrenics, where there is an imbalance of modulating, fountaining chemicals such as dopamine, it might be that excessive release of these powerful agents can suppress the connectivity, particularly in the prefrontal cortex (Ferron, A., Thierry, A. M., Le Douarin, C., and Glowinski, J. [1984] Inhibitory influence of the mesocortical dopaminergic system on spontaneous activity or excitatory response induced from the thalamic mediodorsal nucleus in the rat medial prefrontal cortex. *Brain Research* 302, 257–265. doi:10:1016/0006–8993(84)90238–5. Gao, W. J., Wang, Y., and Goldman-Rakic, P. S. [2003]. Dopamine modulation of perisomatic and peridendritic inhibition in prefrontal cortex. *Journal of Neuroscience* 23, no. 5, 1622–1630. Retrieved from http://neurobio.drexel.edu/GaoWeb/papers/J.%20Neurosci.%202003.pdf), reducing the robustness of cognitive processes and thus enhancing disproportionately the impact of the incoming senses (Greenfield, S.A. [2001]. *The private life of the brain: Emotions, consciousness, and the secret of the self*. New York: Wiley).
28. Tsujimoto, S. (2008). The prefrontal cortex: Functional neural development during early childhood. *The Neuroscientist* 14, 345–358. doi:10:1177/1073858408316002.

29. Shimamura, A. P. (1995). Memory and the prefrontal cortex. *Annals of the New York Academy of Sciences* 769, no. 1, 151–160. doi:10:1111/j.1749–6632:1995 .tb38136.x.

30. Ferron et al., 1984.

31. Lazzaro, N. (2004). *Why we play games: Four keys to more emotion without story*. Retrieved from http://www.xeodesign.com/xeodesign_whyweplaygames .pdf.

32. Fischer, P., Greitemeyer, T., Morton, T., Kastenmüller, A., Postmes, T., Frey, D., . . . and Odenwälder, J. (2009). The racing-game effect: Why do video racing games increase risk-taking inclinations? *Personality and Social Psychology Bulletin* 35, no. 10, 1395–1409. doi:10:1177/0146167209339628.

33. Hull, J. G., Draghici, A. M., and Sargent, J. D. (2012). A longitudinal study of risk-glorifying videogames and reckless driving. *Psychology of Popular Media Culture* 1, no. 4, 244. doi:10:1037/a0029510.

34. Koepp, M. J., Gunn, R. N., Lawrence, A. D., Cunningham, V. J., Dagher, A., Jones, T., . . . and Grasby, P. M. (1998). Evidence for striatal dopamine release during a videogame. *Nature* 393, no. 6682, 266–267. doi:10:1038/30498.

35. Kelly, C. R., Grinband, J., Hirsch, J. (2007). Repeated exposure to media violence is associated with diminished response in an inhibitory frontolimbic network. *PLOS ONE* 2, no. 12, e1268. doi:10:1371/journal.pone.0001268.

36. Yuan, K., Qin, W., Wang, G., Zeng, F., Zhao, L., Yang, X., . . . and Tian, J. (2011). Microstructure abnormalities in adolescents with Internet addiction disorder. *PLOS ONE* 6, no. 6, e20708. doi:10:1371/journal.pone.0001268.

37. A subsequent study looking specifically at gaming addiction showed abnormalities in cortical thickness. Specifically, increased cortical thickness was found in a range of regions, which correlated with the duration of the gaming addiction (Yuan, K., Cheng, P., Dong, T., Bi, Y., Xing, L., Yu, D., . . . and Tian, J. [2013]. Cortical thickness abnormalities in late adolescence with online gaming addiction. *PLOS ONE* 8, no. 1, e53055. doi:10:1371/journal .pone.0053055).

38. Kim, Y. R., Son, J. W., Lee, S. I., Shin, C. J., Kim, S. K., Ju, G., . . . and Ha, T. H. (2012). Abnormal brain activation of adolescent Internet addict in a ball-throwing animation task: Possible neural correlates of disembodiment revealed by fMRI. *Progress in Neuro-Psychopharmacology and Biological Psychiatry* 39, no. 1, 88–95. doi:10:1016/j.pnpbp.2012:05.013.

CHAPTER 16. THE *SOMETHING* ABOUT SURFING

1. Polly, J. (March 22, 2008). Surfing the Internet. Retrieved from http://www .netmom.com/about-net-mom/26-surfing-the-internet.html.

2. But a brief cautionary note: such a utopian state of affairs must be set against the consideration that the Internet is only as useful as the information it disseminates. For example, Steve Bratt, now CEO of the World Wide Web Foundation, gave an insightful explanation for the 3.3-billion-person gap between mobile phone users and Internet users. The problem is that for a person in a developing country, the current Internet is far less useful for everyday life than it might seem: "Maybe they can look at scores from the playoffs, but if they want to find a local doctor, if they want to understand which crops to plant or how much money they can get for their crops, if they want to be able to teach their kids a language other than English or French or Chi-

nese, there's just nothing for them there" (Kessler, S. [February 4, 2011]. Why the Web is useless in developing countries—and how to fix it [blog post]. Retrieved from http://mashable.com/2011/02/04/web-developing-world).

3. Bohannon, J. (December 2011). *Without Google and Wikipedia I am stupid.* Speech given at Online Educa Berlin, Berlin, Germany.

4. Sparrow, B., Liu, J., and Wegner, D. M. (2011). Google effects on memory: Cognitive consequences of having information at our fingertips. *Science* 333, no. 6043, 776–778. doi:10:1126/science.1207745.

5. Squire, L. R., and Zola, S. M. (1996). Structure and function of declarative and nondeclarative memory systems. *Proceedings of the National Academy of Sciences* 93, no. 24, 13515–13522. Retrieved from http://www.ncbi.nlm.nih .gov/pmc/articles/PMC33639.

6. Sparrow, Liu, and Wegner, 2011, p. 776.

7. Small, G. W., Moody, T. D., Siddarth, P., and Bookheimer, S. Y. (2009). Your brain on Google: Patterns of cerebral activation during Internet searching. *American Journal of Geriatric Psychiatry* 17, 116–126. doi:10:1097/JGP.0b013e 3181953a02.

8. Eliot, T. S. (1942). *Little Gidding.* Retrieved from http://www.columbia.edu/ itc/history/winter/w3206/edit/tseliotlittlegidding.html.

9. Thurber, J. (February 18, 1939). Fables for our time—III. *The New Yorker.* Retrieved from http://www.newyorker.com/archive/1939/02/18/1939_02_18 _019_TNY_CARDS_000176433, p.19.

10. Desjarlais, M. (2013). Internet exploration behaviors and recovery from unsuccessful actions differ between learners with high and low levels of attention. *Computers in Human Behavior* 29, no. 3, 694–705. doi:10:1016/ j.chb.2012:12.006.

11. Nicholas, D., Rowlands, I., Clark, D., and Williams, P. (2011). Google Generation II: Web behavior experiments with the BBC. *Aslib Proceedings* 63, no. 1, 28–45. doi:10:1108/00012531111103768, p. 44.

12. In 2011, 86 percent of the U.K. Internet population visited a video site at least once. (Experian Hitwise. [2011]. *Online video: Bringing social media to life.* Retrieved from http://www.experian.co.uk/marketing-services/products/ hitwise.html) In the United States in August 2012, 188 million people viewed an online video, with an average of twenty-two hours' worth of online video viewing per person that month (comScore. [September 19, 2012]. Online video content reaches all-time high of 188 million viewers [blog post]. Retrieved from https://www.comscore.com/esl/Insights/Press_Releases/2012/ 9/comScore_Releases_August_2012_US_Online_Video_Rankings). The majority of online content viewed is through YouTube, which dominates visits to online video sites. In the past few years, YouTube and other sites for sharing video files over the Internet have soared from obscurity to a pivotal position in the media landscape. Networks fear that the availability of their clips so freely and flexibly will depress television viewing, but unauthorized clips are also free advertising for television shows and, as YouTube has grown quickly, major networks have responded by making their content available at their own sites.

13. Waldfogel, J. (2009). Lost on the Web: Does Web distribution stimulate or depress television viewing? *Information Economics and Policy* 21, no. 2, 158– 168. doi:10:1016/j.infoecopol.2008:11.002.

14. Watson, R. (January 28, 2011). Out and about [blog post]. Retrieved from http://toptrends.nowandnext.com/2011/01/28/out-and-about.

CHAPTER 17. THE SCREEN IS THE MESSAGE

1. McLuhan, M. (1994). *Understanding media: The extensions of man.* Cambridge, MA: MIT Press, p. 9.
2. Sellen, A. J., and Harper, R. H. (2001). *The myth of the paperless office.* Cambridge, MA: MIT Press.
3. The study investigated whether there is any difference in comprehension in fifteen- and sixteen-year-olds between reading on a computer screen and in print, and found that students who read in print scored significantly higher on reading comprehension tests than those who read on a screen. A crucial issue was how readily you could get an overview of what was in front of you: readers of the paper text would have immediate access to the text in its entirety. This access, moreover, is built on both visual and tactile cues: the reader can see as well as feel the spatial extension and physical dimensions of the text, as the material substrate of paper provides physical, tactile, and spatiotemporally fixed cues to the length of the text that they are about to read. By contrast, those reading on the screen are restricted to seeing and sensing only one page of the text at any given time. Hence, their overview of the organization, structure, and flow of the text might be hampered. (Mangen, A., Walgermo, B. R., and Brønnick, K. [2013]. Reading linear texts on paper versus computer screen: Effects on reading comprehension. *International Journal of Educational Research* 58, 61–68. doi:10:1016/j.ijer.2012:12.002).
4. Another study of fifth-grade students found that they were more efficient reading from a traditional text than scrolling through a computer screen. The researchers suggested that "difficulties in reading from computers may be due to disrupted mental maps of the text, which may be reflected in poorer understanding and ultimately poorer recall of presented material" (Kerr, M. A., and Symons, S. E. [2006]. Computerized presentation of text: Effects on children's reading of informational material. *Reading and Writing* 19, no. 1, 1–19. doi:10:1007/s11145-003-8128-y).
5. Cataldo, M. G., and Oakhill, J. (2000). Why are poor comprehenders inefficient searchers? An investigation into the effects of text representation and spatial memory on the ability to locate information in text. *Journal of Educational Psychology* 92, no. 4, 791–799. doi:10:1037/0022-0663:92.4:791.

The permanence of words printed on paper helps the reader by providing unequivocal and fixed spatial cues for text memory and recall. In order to respond appropriately to multiple-choice questions, the subjects in Mangen's study were required to locate, access, and retrieve essential pieces of information, either on paper or on screen. Comprehension became harder when the information required to complete such a task, such as answering questions in a reading comprehension assessment, was not immediately visible—for instance, when the reader had to integrate information occurring at locations in a text that were spatially far apart. Such integration requires that the reader has constructed a solid mental representation of the structure of the text. Even with relatively short distances, it is not unreasonable to assume that the fact that the reader cannot touch the digital text in the way he or she touches the pages of a book with the fingers may have challenged the reader's mental reconstruction of the physical layout of the text. In turn, this physical distancing may have impeded the reader's overview as well as his or her ability to access, locate, and retrieve required pieces of textual information.

6. For example, e-book technologies based on electronic ink, such as Kindle and Kobo readers, merely reflect light and are hence more reader-friendly with respect to visual ergonomics, while LCD computer screens cause visual fatigue because they emit light. Various studies have shown that certain features of the LCD screen, such as refresh rate, contrast levels, and fluctuating light, interfere with cognitive processing and hence potentially impair long-term memory, as well as computer vision syndrome, a temporary condition resulting from focusing the eyes on a computer display for protracted, uninterrupted periods of time. Symptoms of this syndrome include headaches, blurred vision, neck pain, redness in the eyes, fatigue, eye strain, dry eyes, irritated eyes, double vision, polyopia, and difficulty refocusing the eyes (Blehm, C., Vishnu, S., Khattak, A., Mitra, S., and Yee, R. M. [2005]. Computer vision syndrome: A review. *Survey of Ophthalmology* 50, no. 3, 253–262. doi:10:1016/j.survophthal.2005:02.008).

7. Jeong, H. (2012). A comparison of the influence of electronic books and paper books on reading comprehension, eye fatigue, and perception. *Electronic Library* 30, no. 3, 390–408. doi:10:1108/02640471211241663.

8. Carr, N. (2011). *The shallows: What the Internet is doing to our brains.* New York: Norton, p. 118.

9. Rideout, V. J., Foehr, U. G., and Roberts, D. F. (2010). *Generation M2: Media in the lives of 8- to 18-year-olds.* Retrieved from http://kaiserfamilyfoundation .wordpress.com/uncategorized/report/generation-m2-media-in-the-lives-of -8-to-18-year-olds.

10. Kessler, S. (2011). 38 percent of college students can't go 10 minutes without tech [blog post]. Retrieved from http://mashable.com/2011/05/31/college -tech-device-stats.

11. Ophir, E., Nass, C., and Wagner, A. D. (2009). Cognitive control in media multitaskers. *Proceedings of the National Academy of Sciences* 106, no. 37, 15583–15587. doi:10:1073/pnas.0903620106.

12. Gorlick, A. (August 24, 2009). Media multitaskers pay mental price, Stanford study shows. Retrieved from http://news.stanford.edu/news/2009/august24/ multitask-research-study-082409.html.

13. Daniel, D. B., and Woody, W. D. (2013). E-textbooks at what cost? Performance and use of electronic v. print texts. *Computers & Education* 62, 18–23. doi:10:1016/j.compedu.2012:10.016.

14. Kraushaar, J. M., and Novak, D. C. (2010). Examining the effects of student multitasking with laptops during the lecture. *Journal of Information Systems Education* 21, 241–251. Retrieved from http://jise.org/Contents/Contents-21 -2.htm.

15. Sana, F., Weston, T., and Cepeda, N. J. (2013). Laptop multitasking hinders classroom learning for both users and nearby peers. *Computers & Education* 62, 24–31. doi:10:1016/j.compedu.2012:10.003.

16. Rosen, L. D., Carrier, L. M., and Cheever, N. A. (2013). Facebook and texting made me do it: Media-induced task-switching while studying. *Computers in Human Behavior* 29, no. 3, 948–958. doi:10:1016/j.chb.2012:12.001.

17. Junco, R., and Cotten, S. R. (2011). Perceived academic effects of instant messaging use. *Computers & Education* 56, no. 2, 370–378. doi:10:1016/ j.compedu.2010:08.020.

18. Rosen, Carrier, and Cheever, 2013. Kirschner, P. A., and Karpinski, A. C. (2010). Facebook and academic performance. *Computers in Human Behavior* 26, no. 6, 1237–1245. doi:10:1016/j.chb.2010:03.024.

19. Kirschner and Karpinski, 2010.
20. Bowman, L. L., Levine, L. E., Waite, B. M., and Gendron, M. (2010). Can students really multitask? An experimental study of instant messaging while reading. *Computers & Education* 54, no. 4, 927–931. doi:10:1016/j.compedu .2009:09.024.
21. A similar kind of study recruited participants who completed a reading comprehension task either uninterrupted or while concurrently holding an instant messaging conversation. Multitaskers took significantly longer to complete the task, indicating that concurrent instant messaging use impaired efficiency. However, while there were no differences between groups in reading comprehension scores, the more time participants reported spending on instant messaging, the lower their reading comprehension scores. Additionally, the more time participants reported spending on instant messaging, the lower their self-reported GPA (Fox, A. B., Rosen, J., and Crawford, M. [2009]. Distractions, distractions: does instant messaging affect college students' performance on a concurrent reading comprehension task? *CyberPsychology & Behavior* 12, no. 1, 51–53. doi:10:1089/cpb.2008:0107).
22. DeStefano, D., and LeFevre, J. A. (2007). Cognitive load in hypertext reading: A review. *Computers in Human Behavior* 23, 1616–1641. doi:10:1016/j.chb.2005 :08.012.
23. Ackerman, R., and Goldsmith, M. (2011). Metacognitive regulation of text learning: On screen versus on paper learning. *Journal of Experimental Psychology: Applied* 17, no. 1, 18–32. doi:10:1037/a0022086, p. 29.
24. Liu, Z. (2005). Reading behavior in the digital environment: Changes in reading behavior over the past ten years. *Journal of Documentation* 61, no. 6, 700–712. doi:10:1016/j.chb.2005:08.012.
25. Kretzschmar, F., Pleimling, D., Hosemann, J., Füssel, S., Bornkessel-Schlesewsky, I., and Schlesewsky, M. (2013). Subjective impressions do not mirror online reading effort: concurrent EEG-eyetracking evidence from the reading of books and digital media. *PLOS ONE* 8, no. 2, e56178. doi:10:1371/ journal.pone.0056178.
26. Daniel, D. B., and Woody, W. D. (2013). E-textbooks at what cost? Performance and use of electronic v. print texts. *Computers & Education* 62, 18–23. doi:10:1016/j.compedu.2012:10.016.
27. Morse, D., Tardival, G. M., and Spicer, J. (1996). *A comparison of the effectiveness of a dichotomous key and a multi-access key to woodlice.* Technical Report 14-96. Retrieved from kar.kent.ac.uk/21343/1/WoodliceMorse.pdf.
28. Farrell, N. (January 7, 2013). Ebook craze is slowing [blog post]. Retrieved from http://news.techeye.net/internet/ebook-craze-is-slowing.
29. Indvik, L. (June 18, 2012). Ebook sales surpass hardcover for the first time in U.S. [blog post]. Retrieved from http://mashable.com/2012/06/17/ebook -hardcover-sales.
30. Flood, A. (February 22, 2013). Decline in independent bookshops continues with 73 closures in 2012. Retrieved from http://www.theguardian.com/ books/2013/feb/22/independent-bookshops-73-closures-2012.
31. Smith, G. E., Housen, P., Yaffe, K., Ruff, R., Kennison, R. F., Mahncke, H. W., and Zelinski, E. M. (2009). A cognitive training program based on principles of brain plasticity: Results from the improvement in memory with plasticity-based adaptive cognitive training study. *Journal of the American Geriatrics Society* 57, 594–603. doi:10:1111/j.1532-5415:2008:02167.x. Thorell, L. B., Lindqvist, S., Nutley, S. B., Bohlin, G., and Klingberg, T. (2009). Training and

transfer effects of executive functions in preschool children. *Developmental Science* 12, 106–113. doi:10:1111/j.1467-7687:2008:00745.x.

32. Owen, A. M., Hampshire, A., Grahn, J. A., Stenton, R., Dajani, S., Burns, A. S., . . . and Ballard, C. G. (2010). Putting brain training to the test. *Nature* 465, no. 7299, 775–778. doi:10:1038/nature09042.

33. Lee, K. M., Jeong, E. J., Park, N., and Ryu, S. (2011). Effects of interactivity in educational games: A mediating role of social presence on learning outcomes. *International Journal of Human-Computer Interaction* 27, no. 7, 620–633. doi:10:1080/10447318:2011:555302.

34. Tanenhaus, J. (n.d.). Computers and special needs: Enhancing self esteem and language. Retrieved from www.kidneeds.com/diagnostic_categories/articles/computers.pdf.

35. Kagohara, D. M., van der Meer, L., Ramdoss, D., O'Reilly, M. F., Lancioni, G. E., Davis, T. N., . . . and Sigafoos, J. (2013). Using iPods and iPads in teaching programs for individuals with developmental disabilities: A systematic review. *Research in Developmental Disabilities* 34, no. 1, 147–156. doi:10:1016/j.ridd.2012:07.027.

36. Li, Q., and Ma, X. (2010). A meta-analysis of the effects of computer technology on school students' mathematics learning. *Educational Psychology Review* 22, 215–243. doi:10:1007/s10648-010-9125-8.

37. Cheung, A. C. K., and Slavin, R. E. (2012). How features of educational technology applications affect student reading outcomes: A meta-analysis. *Educational Research Review* 7, no. 3, 198–215. doi:10:1016/j.edurev.2012:05.002.

38. Gosper, M., Green, D., McNeil, M., Phillips, R., Preston, G., and Woo, K. (2008). *The impact of Web-based lecture technologies on current and future practices in learning and teaching.* Retrieved from http://mq.edu.au/ltc/altc/wblt/docs/report/ce6-22_final2.pdf.

39. Massingham, P., and Herrington, T. (2006). Does attendance matter? An examination of student attitudes, participation, performance, and attendance. *Journal of University Teaching and Learning Practice* 3, no. 2, 82–103. Retrieved from http://ro.uow.edu.au/jutlp/vol3/iss2/3.

40. Brown, B. W., and Liedholm, C. E. (2002). Can Web courses replace the classroom in principles of microeconomics? *American Economic Review Papers and Proceedings* 92, no. 2, 444–448. Retrieved from http://www.jstor.org/stable/3083447.

41. Figlio, D. N., Rush, M., and Yin, L. (2010). *Is it live or is it Internet? Experimental estimates of the effects of online instruction on student learning.* NBER Working Paper No. w16089. Retrieved from http://www.nber.org/papers/w16089.pdf.

42. Korat, O., and Shamir, A. Electronic books versus adult readers: Effects on children's emergent literacy as a function of social class. *Journal of Computer Assisted Learning* 23, no. 3, 248–259. doi:10:1111/j.1365-2729:2006:00213.x.

43. Li and Ma, 2010.

44. Lai, E. (October 16, 2012). Chart: Top 100 iPad rollouts by enterprises & schools [blog post]. Retrieved from http://www.forbes.com/sites/sap/2012/08/31/top-50-ipad-rollouts-by-enterprises-schools.

45. Ramachandran, V. (June 20, 2013). Apple scores $30m iPad deal with L.A. schools [blog post]. Retrieved from http://mashable.com/2013/06/19/la-ipad-tablets-in-schools.

46. Hu, W. (January 4, 2011). Math that moves: Schools embrace the iPad. Re-

trieved from http://www.nytimes.com/2011/01/05/education/05tablets.html
?pagewanted=all&_r=0.

47. Harris, S. (May 29, 2012). The iPad generation: Pupils as young as four taught lessons with a touchscreen. Retrieved from http://www.dailymail.co .uk/news/article-2151403/The-iPad-generation-Pupils-young-taught-lessons -touchscreen.html.

48. Paton, G. (November 12, 2012). Teachers' obsession with technology sees gadgets worth millions sit in cupboards. Retrieved from http://www .telegraph.co.uk/news/9681828/Teachers-obsession-with-technology-sees -gadgets-worth-millions-sit-in-cupboards.html.

49. Furió, D., González-Gancedo, S., Juan, M., Seguí, I., and Rando, N. (2013). Evaluation of learning outcomes using an educational iPhone game vs. traditional game. *Computers & Education* 64, 1–23. doi:10:1016/j.compedu .2012:12.001.

50. Sung, E., and Mayer, R. E. (2013). Online multimedia learning with mobile devices and desktop computers: An experimental test of Clark's methods-not-media hypothesis. *Computers in Human Behavior* 29, no. 3, 639–647. doi:10:1016/j.chb.2012:10.022.

51. Richtel, M. (October 22, 2011). A Silicon Valley school that doesn't compute. Retrieved from http://www.nytimes.com/2011/10/23/technology/at-waldorf -school-in-silicon-valley-technology-can-wait.html?pagewanted=all.

52. Richtel, 2011.

53. Clark, C., and Hawkins, L. (2010). Young people's reading: The importance of the home environment and family support. Retrieved from http://www .literacytrust.org.uk/assets/0000/4954/Young_People_s_Reading_2010.pdf.

54. Topping, K. (2013). What kids are reading: The book-reading habits of students in British schools 2013. Retrieved from http://www.readforpleasure.co .uk/documents/2013wkar_fullreport_lowres.pdf.

CHAPTER 18. THINKING DIFFERENTLY

1. For example, a basic feature of screen-based interaction is the directory tree. This unprecedented constraining of options has become so much a normal part of our lives that we no longer question it, even when we are not sitting directly in front of a screen. Everyone knows the frustration of calling an organization and having our query dealt with not by a real human being but by an automated answering system that inevitably gives us a fixed menu of options with different numbers to press. These directory trees could become the pattern of daily life as we try to gather information, offering us menus with a fixed number of possibilities, where, in order to get to an option that is not immediately presented, we have to plod up and down through various branch lines of categories and subcategories. Given that the human brain becomes good only at what it rehearses, perhaps all that up-and-down to-ing and fro-ing will leave a significant mark on our erstwhile flexible brain processes and, in particular, how we approach problems. Such a rigid and systematic strategy may, on one hand, impart a certain rigor and logic to our debating ability but, on the other hand, turn out to be highly restrictive.

2. There are many different kinds of IQ tests using a wide variety of methods.

Some are visual, some verbal; some use abstract reasoning problems, and still others concentrate on arithmetic, spatial imagery, reading, vocabulary, memory, or general knowledge. Modern comprehensive IQ tests no longer give a single score. Although they still provide an overall rating, they also give values for different abilities, identifying particular strengths and weaknesses of an individual's brainpower.

3. Kurzweil, R. (2005). *The singularity is near: When humans transcend biology.* New York: Penguin.

4. Benyamin, B., Pourcain, B., Davis, O. S., Davies, G., Hansell, N. K., Brion, M. J., . . . and Visscher, P. M. (2013). Childhood intelligence is heritable, highly polygenic and associated with FNBP1L. *Molecular Psychiatry* 19, no. 2, 253–258. doi:10:1038/mp.2012:184.

5. This figure fits well with the calculation of Roger Gosden in his book *Designing Babies,* where he gives an estimate of 0.3 for inheritability of IQ, where a score of 0 indicates a trait attributable entirely to environmental factors and 1 a trait attributable just to genes. However, the crucial issue is that even the inherited component of IQ is the result of many different and diverse genes. If a major gene for IQ were suddenly to be discovered, it would only account for some 5 percent at most and therefore boost an IQ score of 100 to a paltry 101.5 (Gosden, R. G. [1999]. *Designing babies: The brave new world of reproductive technology.* New York: Freeman).

6. Flynn, J. R. (2006). The Flynn effect: Rethinking intelligence and what affects it. In *Introduction to the Psychology of Individual Differences.* Porto Alegre, Brazil: ArtMed.

7. Johnson, S. (2006). *Everything bad is good for you: How today's popular culture is actually making us smarter.* New York: Penguin.

8. Pernille Olesen and his colleagues at the Karolinska Institute in Sweden investigated the changes in brain activity induced by working memory training. Experiments were carried out in which healthy adult human subjects practiced working memory tasks for five weeks. Brain activity was measured with functional magnetic resonance imaging (fMRI) before, during, and after training. After training, brain activity that was related to working memory had indeed increased in key parts of the outer layer of the brain, the cortex (Olesen, P. J., Westerberg, H., and Klingberg, T. [2003]. Increased prefrontal and parietal activity after training of working memory. *Nature Neuroscience* 7, no. 1, 75–79. doi:10:1038/nn1165).

9. Johnson, 2006, p. 39.

10. Paton, G. (October 16, 2010). Facebook "encourages children to spread gossip and insults." Retrieved from http://www.telegraph.co.uk/education/educationnews/8067093/Facebook-encourages-children-to-spread-gossip-and-insults.html.

11. Dickens, C. (2007). *Hard Times.* New York: Pocket Books, p. 1.

12. This notion of real understanding as opposed to information processing fits well with the distinction of "crystallized" as opposed to "fluid" intelligence, the terms first developed by Raymond Cattell (see Chapter 7) (Cattell, R. B. [Ed.] [1987]. *Intelligence: Its structure, growth and action.* Amsterdam: Elsevier).

13. Flynn, 2006.

14. Flynn, J. R. (1987). Massive IQ gains in 14 nations: What IQ tests really measure. *Psychological Bulletin* 101, 171–191. Retrieved from http://www.iapsych.com/iqmr/fe/LinkedDocuments/flynn1987.pdf. Flynn, J. R. (1994). *IQ gains over time.* In R. J. Sternberg (Ed.), *Encyclopedia of human intelligence,* pp. 617–

623. New York: Macmillan. Geake, J. G. (2008). The neurobiology of giftedness. Retrieved from http://hkage.org.hk/b5/events/080714%20APCG/01
-%20Keynotes%20%26%20Invited%20Addresses/1:6%20Geake_The%20
Neurobiology%20of%20Giftedness.pdf.

15. Flynn, 1994.

16. Flynn, 1994.

17. Song, M., Zhou, Y., Li, J., Liu, Y., Tian, L., Yu, C., and Jiang, T. (2008). Brain
spontaneous functional connectivity and intelligence. *Neuroimage* 41, no. 3,
1168–1176. doi:10:1016/j.neuroimage.2008:02.036.

18. Janowsky, J. S., Shimamura, A. P., and Squire, L. R. (1989). Source memory
impairment in patients with frontal lobe lesions. *Neuropsychologia* 27, no. 8,
1043–1056. doi:10:1016/0028-3932(89)90184-X.

19. Gabriel, S., and Young, A. F. (2011). Becoming a vampire without being bit-
ten: The narrative collective-assimilation hypothesis. *Psychological Science*
22, no. 8, 990–994. doi:10:1177/0956797611415541, p. 993.

20. Quoted in Flood, A. (September 7, 2011). Reading fiction "improves empathy" study finds. *Guardian*. Retrieved from http://www.theguardian.com/
books/2011/sep/07/reading fiction empathy-study.

21. Macintyre, B. (November 5, 2009). The Internet is killing storytelling. Retrieved from http://www.thetimes.co.uk/tto/opinion/columnists/benmacintyre/
article2044914.ece.

22. Schonfeld, E. (March 7, 2009). Eric Schmidt tells Charlie Rose Google is "unlikely" to buy Twitter and wants to turn phones into TVs [blog post]. Retrieved from http://techcrunch.com/2009/03/07/eric-schmidt-tells-charlie
-rose-google-is-unlikely-to-buy-twitter-and-wants-to-turn-phones-into-tvs.

23. Sass, L. (2001). Schizophrenia, modernism, and the "creative imagination":
On creativity and psychopathology. *Creativity Research* 13, no. 1, 55–74.
doi:10:1207/S15326934CRJ1301_7.

24. Burnet's clonal selection hypothesis explains how the immune system combats infection by selecting certain types of B and T lymphocytes for the de-
struction of specific antigens invading the body.

25. van Rheenen, E. (January 2, 2014). 12 predictions Isaac Asimov made about
2014 in 1964. Retrieved from http://mentalfloss.com/article/54343/12
-predictions-isaac-asimov-made-about-2014-1964.

CHAPTER 19. MIND CHANGE BEYOND THE SCREEN

1. Office for National Statistics. (2012). *What are the chances of surviving to age
100?* Retrieved from http://www.ons.gov.uk/ons/dcp171776_260525.pdf.

2. While there may not be single genes for complex mental traits, there are specific genes and genetic profiles for certain diseases such as cystic fibrosis and
the degenerative neurological disorder, Huntington's disease. As we saw in
Chapter 5, genes don't work in isolation but need the context of the brain
and body and the environment to manifest, but the fact remains that a
faulty gene will occasionally lead directly, or most often indirectly, to some
kind of correspondingly faulty outcome. So although genes are usually not
the whole story, the detection of a malfunctioning gene can have significant
health benefits. Genetically based technologies therefore hold out the very
real promise of bespoke medication, fitted to an individual's particular genetic profile so that side effects can be minimized (pharmacogenomics).

Even better, an individual's particular risk factors for certain diseases can be predicted by screening the person's genome for preventive measures (lifestyle, diet, or treatment), which can then start as soon as possible (What is pharmacogenomics? [August 12, 2013]. Retrieved from http://ghr.nlm.nih .gov/handbook/genomicresearch/pharmacogenomics).

3. *Regenerative medicine* is a therapy that offers another exciting and realistic alternative, as well as providing a very valuable tool to gain a better understanding of the diseases themselves. The rationale is completely different from that of conventional treatments. The idea is not to treat the symptoms but to harness biological mechanisms that convert basic all-purpose stem cells into specialized cells. These stem cells are injected into the region of the body where the specialized types of cells are deficient. Stem cells, derived from the microscopic ball of some two hundred cells that make up the early-stage embryo, are extraordinary because they have the capacity to produce every single type of cell in the body, whether of the heart, bone, or even brain. This regenerative strategy does not try to compensate for the aberrant effects of a disease as a result of cell death, but from the get-go it actually provides substitutes, generating new cells and supporting ailing cells with the natural chemicals produced by the newcomers. In this way, stem cell therapy brings us that much closer to a real cure for a whole range of diseases, such as the degenerative movement disorder Parkinson's disease, but not for Alzheimer's disease, where the sites of damage are too widespread (Pera, R. A. R., and Gleeson, J. G. [2008]. Stem cells and regeneration [special review issue]. *Human Molecular Genetics, 17*(R1), R1–R2. doi:10:1093/hmg/ddn186).

4. Knapp, M., and Prince, M. (2007). *Dementia UK*. Retrieved from http://www .alzheimers.org.uk/site/scripts/download.php?fileID=1.

5. Prince, M., Bryce, R., Albanese, E., Wimo, A., Ribeiro, W., and Ferri, C. P. (2013). The global prevalence of dementia: A systematic review and meta-analysis. *Alzheimer's & Dementia* 9, no. 1, 63–75. Retrieved from http://www .sciencedirect.com/science/article/pii/S1552526012025319.

6. Hurd, M. D., Martorell, P., Delavande, A., Mullen, K. J., and Langa, K. M. (2013). Monetary costs of dementia in the United States. *New England Journal of Medicine* 368, no. 14, 1326–1334. doi:10:1056/NEJMsa1204629.

7. Atkins, S. (2013). *First steps in living with dementia*. Oxford: Lion Hudson.

8. Currently, the diagnosis of Alzheimer's disease is based provisionally on symptoms, and ultimately on postmortem histological verification. By the time the patient presents with the characteristic cognitive impairments, extensive atrophy of the affected brain regions has typically already been underway for some ten or even twenty years. Arguably, there has been no dramatically different new drug introduced to combat Alzheimer's disease specifically, or neurodegenerative disorders more generally since 1996, when approval from the U.S. Food and Drug Administration (FDA) was granted for donepezil under the brand name Aricept. The reason is that there is as yet no accepted or proven basic mechanism that could consequently be targeted pharmaceutically. One well-established idea is that the key problem is a deficit of a certain chemical messenger, the neurotransmitter acetylcholine, as a result of the death of specific neurons where it is operational (Bartus, R. T., Dean, R. L., Pontecorvo, M. J., and Flicker, C. [1985].) The cholinergic hypothesis: A historical overview, current perspective, and future directions.

Annals of the New York Academy of Sciences 444, no. 1, 332–358. doi:10:1111/
j.1749-6632:1985.tb37600.x. Terry, A. V., and Buccafusco, J. J. [2003]. The
cholinergic hypothesis of age and Alzheimer's disease–related cognitive deficits: Recent challenges and their implications for novel drug development.
Journal of Pharmacology and Experimental Therapeutics 306, no. 3, 821–827).
Hence the treatment of choice at the moment is Aricept (or an equivalent
such as galantamine, sold, for example, as Reminyl), a drug that temporarily
boosts levels of the dwindling neurotransmitter acetylcholine by protecting it
from normal enzymatic degradation. However, this theory fails to account for
a well-known discrepancy: not all areas of the brain affected by Alzheimer's
use acetylcholine, nor are all acetylcholine-using areas of the brain affected by
the disease. Not surprisingly, therefore, Aricept does not prevent the continuing death of cells, since it merely tackles just one biochemical symptom.

The other main contender for accounting for the process of neurodegeneration is the *amyloid hypothesis,* where neuronal death is attributed to disruption of the cell structure by toxic deposits of a substance named amyloid
(after the Greek word for "starch"), which is characteristic of the postmortem Alzheimer brain (Hardy, J., and Allsop, D. [1991]. Amyloid deposition as
the central event in the etiology of Alzheimer's disease. *Trends in Pharmacological Sciences* 12, 383–388. doi:10:1016/0165-6147(91)90609-V. Hardy, J. A.,
and Higgins, G. A. [1992]. Alzheimer's disease: The amyloid cascade hypothesis. *Science* 256, no. 5054, 184–185. doi:10:1126/science.1566067. Pákáski,
M., and Kálmán, J. [2008]. Interactions between the amyloid and cholinergic mechanisms in Alzheimer's disease. *Neurochemistry International* 53, no.
5, 103–111. doi:10:1016/j.neuint.2008:06.005). However, the amyloid hypothesis does not explain the fact that only certain cells are vulnerable in
neurodegenerative diseases, the absence of amyloid deposits in some otherwise faithful animal models of dementia, or indeed the occurrence postmortem of amyloid in the healthy non-Alzheimer brain. Again, it is small wonder
that, despite the popularity of amyloid formation as a pharmaceutical target
since the 1990s, no treatment based on this theory has yet proved effective
in the clinic.

9. Of course, many neuroscientists, including my own group, have their own
   ideas as to what this crucial, pivotal neurodegenerative process may be. For a
   technical review of our group research on dementia, see Greenfield, S.
   (2013). Discovering and targeting the basic mechanism of neurodegeneration: The role of peptides from the C-terminus of acetylcholinesterase.
   Chemico-biological Interactions 203, no. 3, 543–546. doi:10:1016/j.cbi.2013:03
   .015.

10. Baker, R. (1999). *Sex in the future: Ancient urges meet future technology.* London:
    Macmillan.

11. Rosegrant, S. (n.d.). The new retirement: No retirement? [blog post]. Re-
    trieved from http://home.isr.umich.edu/sampler/the-new-retirement.

12. Forecasts from *The Futurist* magazine. (n.d.). Retrieved from http://www.wfs
    .org/Forecasts_From_The_Futurist_Magazine.

13. Glass. (n.d.). Retrieved from http://www.google.com/glass/start.

14. The actual term "augmented reality" is believed to have been coined by Pro-
    fessor Tom Caudell, working for Boeing, and first used in 1990. Since then a
    variety of applications of increasing sophistication have been developed,
    culminating in Google Glass (Sung, D. [March 1, 2011]. The history of aug-

mented reality [blog post]. Retrieved from http://www.pocket-lint.com/news/108888-the-history-of-augmented-reality).

15. Graham, M., Zook, M., and Boulton, A. (2012). Augmented reality in urban places: Contested content and the duplicity of code. *Transactions of the Institute of British Geographers* 38, no. 3, 464–479. doi:10:1111/j.1475-5661:2012 :00539.x.

16. ABI Research. (February 21, 2013). Wearable computing devices, like Apple's iWatch, will exceed 485 million annual shipments by 2018. Retrieved from https://www.abiresearch.com/press/wearable-computing-devices-like -apples-iwatch-will.

17. Lookout. (2012). *Mobile mindset study*. Retrieved from https://www.lookout .com/resources/reports/mobile-mindset.

18. Securenvoy. (February 16, 2012). 66 percent of the population suffer from nomophobia: The fear of being without their phone [blog post]. Retrieved from   http://www.securenvoy.com/blog/2012/02/16/66-of-the-population -suffer-from-nomophobia-the-fear-of-being-without-their-phone.

19. Keen, A. (February 26, 2013). Why life through Google Glass should be for our eyes only [blog post]. Retrieved from http://edition.cnn.com/2013/02/ 25/tech/innovation/google-glass-privacy-andrew-keen.

20. Haworth, A. (October 20, 2013). Why have young people in Japan stopped having sex? Retrieved from http://www.theguardian.com/world/2013/oct/ 20/young-people-japan-stopped-having-sex. Low birthrate could slash South Korea's youth population in half by 2060: Report. (January 8, 2013). Retrieved from http://www.japantimes.co.jp/news/2013/01/08/asia-pacific/ low-birthrate-could-slash-south-koreas-youth-population-in-half-by-2060 -report.

21. Haworth, 2013.

22. Haworth, 2013.

23. Trends in obesity prevalence. (n.d.). Retrieved from http://www.noo.org.uk/ NOO_about_obesity/trends.

CHAPTER 20. MAKING CONNECTIONS

1. Wang, Y. (March 25, 2013). More people have cell phones than toilets, U.N. study shows. Retrieved from http://newsfeed.time.com/2013/03/25/more -people-have-cell-phones-than-toilets-u-n-study-shows/#ixzz2cEZUrSIF.

2. Frequently asked questions. (April 10, 2007). Retrieved from http://www-03 .ibm.com/ibm/history/documents/pdf/faq.pdf.

3. Haldane, J. B. S. (1923). *Daedalus: Or, science and the future*. London: Kegan Paul. Retrieved from http://vserver1.cscs.lsa.umich.edu/~crshalizi/Daedalus .html.

4. Haldane, 1923.

5. Quoted in Apt46. (May 18, 2011). Socrates was against writing [blog post]. Retrieved from http://apt46.net/2011/05/18/socrates-was-against-writing.

6. Muir, E. (1952). The horses. Retrieved from http://www.poemhunter.com/ best-poems/edwin-muir/the-horses.

7. Bettelheim, B. (1959). Joey: A "mechanical boy." Retrieved from http://www .weber.edu/wsuimages/psychology/FacultySites/Horvat/Joey.PDF.

8. Oxford dictionaries word of the year. (November 19, 2013). Retrieved from

http://blog.oxforddictionaries.com/press-releases/oxford-dictionaries-word
-of-the-year-2013.

9. James, O. (2008). *Affluenza*. London: Vermilion.

10. Russell, B. (1924). Icarus or the future of science. Retrieved from http://www
.marxists.org/reference/subject/philosophy/works/en/russell2.htm.

11. Criado-Perez, C. (August 7, 2013). Diary: Internet trolls, Twitter rape threats
and putting Jane Austen on our banknotes. Retrieved from http://www
.newstatesman.com/2013/08/internet-trolls-twitter-rape-threats-and-putting
-jane-austen-our-banknotes.

12. Zimmer, B. (April 26, 2012). What is YOLO? Only teenagers know for sure.
Retrieved from http://www.bostonglobe.com/ideas/2012/08/25/what-yolo
-only-teenagers-know-for-sure/Idso04FecrYzLa4KOOYpXO/story.html.

13. I've previously dubbed the three basic tendencies for self-expression *someone*
(the drive for status), where you are individual but not fulfilled; *anyone* (the
appeal of the collective identity as previously exemplified in various political movements such as fascism and communism); and *nobody* (the need for
abnegating the sense of self and living in/for the moment). See Greenfield, S.
(2008). *I.D.: The quest for meaning in the 21st century.* London: Hodder &
Stoughton.

14. Examples of such software might include:

Other people's minds. The aim here would be to combat problems in empathy. The experience would start with a conventional visual sequence of fast-
moving events, driven by the user. The speed of the images would be slowed
incrementally, with longer periods introduced for speech, then conversation. Note that it would be valuable for such software to use voices with different inflections, re-creating the experience of prosody. Questions would
be inserted intermittently, querying the various outcomes that might potentially result from what the different people in the ongoing scene might
do, and progressing according to what these suggest. Previous performance
would set the skill level for empathy. What does it all mean? Building up over
time, this would become an individual's conceptual framework. The user enters random ideas—brainstorming or, indeed, as if blogging, interesting
facts learned, even titles of books read. An individual framework would be
developed that then feeds into other responses/activities; e.g. the notion
"the government is betraying us" might be cross-referenced with other examples within the existing personal framework, then to a wider, more objective database. Evaluations would show progress based on the understanding
of abstract ideas, but from an individual perspective.

Consequences. The idea here would be to reinforce the message that, after
all, actions really do have consequences. It would consist of a suite of games
in which permanent change results from action: for example, if someone is
shot dead, they remain dead thereafter. For every action, such as being shot,
the program would cut to real-life footage that includes a brief report from
someone on what it actually feels like to be shot or bereaved, for example.

Imagine. The idea here would be to tackle the constraints imposed by anything from PowerPoint to word processing to company answering machine
messages, all suggesting that life has only a fixed number of options. So, no
menus! The starting point is a word/idea/action of the user's own that freely
links to anything else and is prompted by previous entries across the whole
range of other programs. Icons/pictures of the entry are slowly replaced by

words/voice. Over time, the entries build up into an increasingly complex, evolving conceptual framework.

My life story. The aim is to reclaim a sense of privacy. It would be a Facebook-type activity, but for the user only, and locked into real time. Since this "diary" will be impossible to share with anyone else, the user will develop a sense of privacy and an enduring sense of self with a clear narrative that the user learns does not require feedback or comments from others. It may be best appreciated as a smartphone app, in which the user could confide wherever and whenever they wished.

Who am I? This would be an attempt to bolster up a sense of identity. Here there would be feedback to the user based on the user's inputs over time. As performance on the whole range of activities accumulates, an analysis of the responses builds up of the types of personality traits emerging and/or changing.

FURTHER READING

Atkins, S. (2013). *First steps to living with dementia.* Oxford: Lion.

Baker, R. (1999). *Sex in the future: Ancient urges meet future technology.* London: Macmillan.

Begley, S. (2008). *The plastic mind.* London: Constable & Robinson.

Bowlby, J. (1969). *Attachment and loss,* Vol. 1: *Loss.* New York: Basic Books.

Carr, N. (2011). *The shallows: What the Internet is doing to our brains.* New York: Norton.

Flynn, J. R. (2006). The Flynn Effect: Rethinking intelligence and what affects it. In *Introduction to the Psychology of Individual Differences.* Porto Alegre, Brazil: ArtMed.

Frost, J. L. (2010). *A history of children's play and play environments: Toward a contemporary child-saving movement.* New York: Routledge

Gentile, D. A. (Ed.) (2003). *Media violence and children: A complete guide for parents and professionals.* Westport, CT: Praeger.

Gosden, R. G. (1999). *Designing babies: The brave new world of reproductive technology.* New York: Freeman.

Greenfield, S. (2011). *You and me: The neuroscience of identity.* London: Notting Hill.

Greenfield, S. A. (2001). *The private life of the brain: Emotions, consciousness, and the secret of the self.* New York: Wiley.

Greenfield, S. A. (2007). *I.D.: The quest for meaning in the 21st Century.* London: Hodder.

Harkaway, N. (2012). *The blind giant: Being human in a digital world.* London: Vintage.

Johnson, S. (2006). *Everything bad is good for you: How today's popular culture is actually making us smarter.* New York: Penguin.

Keen, A. (2007). *The cult of the amateur.* London: Nicholas Brealey.

Kurzweil, R. (2005). *The singularity is near: When humans transcend biology.* New York: Penguin.

McLuhan, M. (1994). *Understanding media: The extensions of man.* Cambridge, MA: MIT Press.

Palmer, S. (2007). *Toxic childhood: How the modern world is damaging our children and what we can do about it.* London: Orion.

Pickren, W., and Rutherford, A. (2010). *A history of modern psychology in context.* Hoboken, NJ: Wiley.

Purves, D., Augustine, G. J., Fitzpatrick, D., Hall, W. C., LaMantia, A. S., and White, L. E. (Eds.) (2012). *Neuroscience* (5th ed.). Sunderland, MA: Sinauer.

Rosen, L. D. (2012). *iDisorder: Understanding our obsession with technology and overcoming its hold on us.* New York: Macmillan

Sellen, A. J., and Harper, R. H. (2003). *The myth of the paperless office.* Cambridge, MA: MIT Press.

Turkle, S. (2011). *Alone together: Why we expect more from technology and less from each other.* New York: Basic Books.

Watson, R. (2010). *Future files: A brief history of the next 50 years.* London: Nicholas Brealey.

INDEX

ABOUT THE AUTHOR

Baroness Susan Greenfield, CBE, FRCP (Hon) was both an undergraduate and graduate at Oxford University, earning a DPhil in the Department of Pharmacology in 1977. She subsequently held research fellowships in the Department of Physiology, Anatomy, and Genetics, Oxford; the Collège de France, Paris; and NYU Langone Medical Center, New York. From 1998 to 2010 she served as director of the Royal Institution of Great Britain, a post held jointly with her chair in Oxford. She currently holds a senior research fellowship at Oxford University, Lincoln College, and is CEO/CSO of a biotech company (www.neuro-bio.com) that is developing a novel anti-Alzheimer's drug based on her research exploring brain mechanisms linked to neurodegeneration.

Greenfield has since been awarded thirty-one honorary degrees from British and foreign universities. In 2000 she was elected to an honorary fellowship of the Royal College of Physicians. Further international recognition of her work has included the Golden Plate Award (2003) from the Academy of Achievement, Washington; the Ordre national de la Légion d'honneur (2003) from the French Government; and the 2010 Australian Society for Medical Research medal. She was awarded a CBE in the Millennium New Year Honours list, and was granted a non-political life peerage in 2001. In 2004 and 2005, she was Thinker in Residence in Adelaide, reporting to the Premier of South Australia on applications of science for wealth creation. She was appointed chancellor of Heriot-Watt University, 2005–2012, and in 2007 she was elected into the fellowship of the Royal Society of Edinburgh.

She has recently completed a visiting professorship (November 2014) at the Medical School, University of Melbourne, Australia.

Due to her original training in classics, Greenfield held the presidency of the Classical Association for 2003–2004 and in 2010 was elected to a fellowship of the Science Museum. She was a Forum Fellow at the World Economic Forum in Davos for ten years. In response to a request in 2002 from the secretary of state for Trade and Industry, she authored *Set Fair: A Report on Women in Science, Engineering, and Technology*. Greenfield has been profiled in a wide range of papers and magazines, voted one of the 100 most influential women in Britain by the *Daily Mail* in 2003, and Woman of the Year by the *Observer* in 2000. Greenfield has also been selected in Britain's 500 most Influential People by Debrett's.

www.susangreenfield.com

ABOUT THE TYPE

This book was set in Stone, a typeface created in 1988 by the type and graphic designer Sumner Stone (b. 1945). This typeface was designed to satisfy the requirements of low-resolution laser printing. Its traditional design blends harmoniously with many typefaces, making it appropriate for a variety of applications.